"Heath is a very special thing. In our rapidly changing world, it is a company with human values, dedicated to quality in all that it does. There are lessons to be learned from its approach. At Heath's heart is a wonderful product. I've always said that materials carry hidden messages and, for anyone who struggles to understand that, I recommend a trip to Heath to feel the power of ceramics."

—Ilse Crawford, designer

"Heath is responsible for the most beautiful ceramics in America, and perhaps the world. Cathy and Robin not only make tile, but their whole lives are also about tile. Nobody knows it better and nobody looks at it with the same appreciation and love. This book is an expression of their curiosity and infinite knowledge for the subject, plus it's the only reference book of its kind."

—Roman Alonso, Commune Design

TILE MAKES THE ROOM

TILE MAKES THE ROOM

GOOD DESIGN FROM HEATH CERAMICS

Catherine Bailey & Robin Petravic

TEN SPEED PRESS
BERKELEY

Editorial Direction
Simone Silverstein

Photography Direction
Mariko Reed

Design
Volume Inc.

 Published in the United States by Ten Speed Press, an imprint of the Crown Publishing Group, a division of Penguin Random House LLC, New York.
www.crownpublishing.com
www.tenspeed.com

Library of Congress Cataloging-in-Publication Data
Bailey, Cathy, author.
Tile makes the room / Cathy Bailey & Robin Petravic.
pages cm
1. Heath Ceramics. 2. Ceramic tiles—California—Sausalito.
I. Petravic, Robin, author. II. Title.
NK4670.7.U53H433 2015
738.6—dc23
2015013735

Hardcover ISBN: 978-1-60774-741-3
eBook ISBN: 978-1-60774-742-0

Printed in China

10 9 8 7 6 5 4 3 2 1
First Edition

CONTENTS

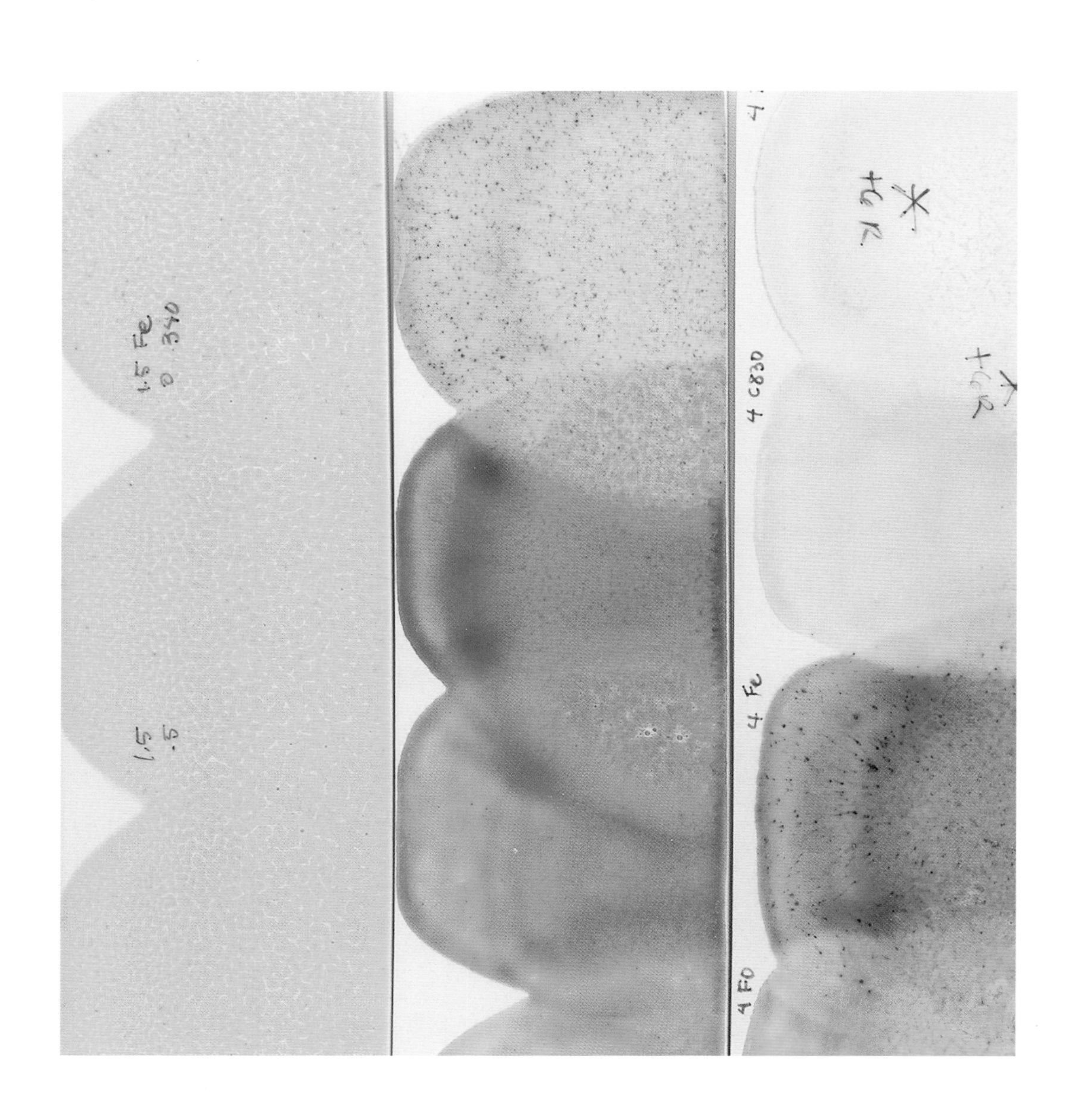
1.5 Fe
4 C830
4 Fe
4 FO

PREFACE: BUILDING ENVIRONMENTS AROUND OUR LIVES

We create the spaces we inhabit around the lives we lead. From work to home and to work again there's commonality, though they may feel different at first glance. We've designed these spaces based on how we spend our time and how we want to feel when we work, relax, and entertain in them. Each space creates a backdrop to our lives, inspiring us and reminding us of where we've been and what we care about.

The way we use environments changes as our children grow, priorities shift, as well as how we work and spend our time. Nothing should be too precious or static—this is reason enough to lay a foundation with good, honest materials. We find that when the materials chosen respect the space's purpose, it's beautiful, and obvious. Nothing needs to be too expensive or rare, but it must have intention, the ability to gain a patina over time and to tell a story.

Good design (for us) is not a particular style. It's respect for what truly works in a space—be it form or function, how it enhances our experience, and how it evolves over time. There's no old and new, only what moves us forward as we live our lives.

HEATH
200
400

D3
D2
Universal Strap
UNIVERSAL banding
Bud Vase
MC2
Medium Vase
MC3
119 M
20 1
TEAPOT LARGE

OUR 1959 SAUSALITO FACTORY

This midcentury building is steeped in heritage; around every corner the past collides with the present. The history of the building, with its roots as a small pottery, inspires us as we move forward. What works? What can we leave alone? What, if anything, needs to be changed? Being in this factory reminds us of simpler times. A typewriter sits on a shelf beside Robin's desk, sharing the space with electrical outlets charging all the gadgets we use today. In the factory, windows line work-stations, providing natural light and keeping the environment human and humane. Every surface is splattered in all manner of clay and glaze. Rather than cleaning them up, we witness ever-changing works of art that pay tribute to the making of ceramics. The ceiling's folded plywood trusses, with long open spans, make for a lightweight feel—a young architect's experiment that never caught on—providing a strong and unique design element. Since it's next to the bay, the building is prone to flooding when tides are high. As luck would have it, the concrete floors wipe clean and dry with a simple mopping. As we update and adapt the building, we make choices that work around the functionality and materials from its original design. The building creates thoughtful conversation—working in it is living in it.

AMACO
3M
VWR
Hertz

A

OUR 1890s HOME

We chose our "Victorian" farmhouse for its location and proportions. We call it Victorian not because of its ornate details, but because its narrow windows and ten-foot ceilings remind us of that architectural style. We've carefully rebuilt the house over time, keeping the proportions and coziness intact on each of its three floors. We wanted the former basement to feel warm—our sanctuary for lazy movies and long baths. We tiled the guest bedroom in an olive green that flows into chocolate brown tile in the family room that flows into meadow green tile in the bathroom. The main floor, where daily life happens, was opened up to connect the kitchen, dining, and living rooms. Six of us live here (three humans, three animals) and we're often in separate rooms, but each is close enough to enjoy each other's conversation and the flow of everyday life. The home is lined with simple Victorian moldings, tall windows that complement tall ceilings, and natural fir floors in a mix of old and new planking. Black, ornately patterned Josef Frank wallpaper in the living room gives way to the inverse of the same wallpaper pattern, in white, going up the staircase to the bedrooms. Upstairs is light and warm, a place for dozing off and preparing for the day ahead. A former attic, the white painted beadboard on the walls and ceiling make the bedrooms feel larger than they are and frame views of the bay. Original soft fir floors age and wear next to the new soft fir planks beside them.

THE CROSSING
THINGS I HAVE LEARNED

HANDCRAFTED MODERN

79

OUR 2012 SAN FRANCISCO FACTORY

This building was originally a commercial laundry, built in an industrial neighborhood of San Francisco. Today it houses Heath Ceramics—a tile factory, a showroom, and our creative studios where we design, manufacture and sell under one roof. Before we moved in, the windows had been painted over. There were many small, broken-up spaces limiting functionality. Not only have we uncovered and replaced nearly every window, we've added more—natural light flowing in keeps the space feeling more humane. Our upstairs design studios get their light from tall steel frame windows, while the factory itself gets light from a four-sided clerestory held up by heavy wooden timbers. The factory also benefits from the floor-to-ceiling glass walls that we added to separate it from the showroom, the factory, public viewing court and café. This transparency is a key architectural element, connecting the making of our tile with the selling of our tile, which creates a more democratic relationship between the people that make and the people that buy. It means that every element of our factory is as thoughtfully designed as our studios. We've created a building that blurs typical lines of division and houses a distinct community—designers, makers, retailers, shoppers, diners, and observers—all under the roof of a tile factory.

EXIT

EXIT

6
AMERICAN RANGE

heath
OPTIMAL
CE

HEATH
CAUTION !!!
HOT SURFACE

INTRODUCTION

We don't think of Heath as making tile so much as making objects that contribute to something bigger, like architecture and interiors.

We love tile—its shape, color, texture, and pattern. Its functional and decorative characteristics are as relevant today as they were thousands of years ago. We especially love when a room calls for a combination of materials and tile is one of them, because it mixes so beautifully with other materials. It's versatile in its form, function, and aesthetic. A hard surface to the touch becomes organic, soft, grounding, and adaptable when paired with opposite or complementary materials like textiles, woods, or glass. It molds and morphs and fits in such a wide variety of inspiring ways. Without a doubt, every time we see a tile installation that "works," it's because it's cohesive with the overall design.

This book is about contemporary spaces (though good design always nods to the past) that use tile in creative ways. It's about a tile installation's ability to drive design, and the questions one ought to ask when working with an element, like tile, that has limitless expression and opportunity. It's about our love for the material and what we've learned from making it at Heath. We don't see tile as an individual piece in a particular size or color. We see the installation that's brought to life, like a picture on a wall, or cladding a building or floor—an expression of the space and of the mood you wish to create. This way of thinking about tile takes it beyond a building material, one you simply buy off the shelf, and elevates it to an art or craft.

Before we go further, a bit of context. We've been known to talk through the houses we've lived in since birth and catalog the design elements we remember about each. It turns out, popular materials used in the era we grew up were materials that we watched wear out in a short period of time—linoleum floors, flocked wallpapers, wall-to-wall carpeting, to name a few. Needless to say, neither of us lived in houses with inspiring use of tile, so our interest came from a broader appreciation of design.

For Robin, a modest brick house in England (where he was born) instilled strong memories of thick cork floors. His next move was to New Jersey, to a contemporary house his parents built among a neighborhood of Colonials. The house had wonderfully detailed mahogany stair rails and doors, but no tile of note, only dull bathroom colors. Then there was the stone cottage his grandfather built on the Croatian coast for family gatherings by the sea. In a perpetual state of construction, in character with its hot seaside setting, and seemingly the only finished element being its tiled terracotta roof. It had a permanent exterior of red hollow brick (the perfect lizard hideout), but remained without its stucco finish. There were building materials sitting in piles and the air always smelled of freshly mixed concrete, prompting one to wonder if the sandbox was for the kids or the stonemasons.

Also in New Jersey, Cathy's split-level Colonial was built in the era of pink tile and mushroom brown appliances. Their weekend home on a lake in Pennsylvania began as a wooden shell and evolved weekend after weekend at the hands of its family members—Cathy and her mother tiled a floor in a non-descript tan color, though they've both become much more discerning since. (Note the common thread of DIY and family vacations.) Her father worked in a very cool, memorable building, which was the headquarters of Bell Labs. She discovered later that the building was designed by Eero Saarinen. Made of glass and concrete, materials rarely used in such pure combinations by architects in the Garden State suburbs where we grew up.

We've both grown to truly appreciate materials that last. A lot of our interest comes from old buildings, historical structures that, because of their materials and how they aged, emanate respect from the inside out. Through travel, we've experienced grand places like the Roman Forum, the Alhambra Palace, Venetian palazzos, old European churches, Mexican colonial cities, and train stations like Grand Central (but not Penn!), and houses from Bauhaus masters like Walter Gropius or craftsman like J.B. Blunk, and even Swiss farmhouses made by an unknown builder of stone and roughly hewn wood. There's an intangible feeling one gets from a structure built with the intent on being around longer than any one resident. These are the environments that we call upon for inspiration or when viewing spaces we haven't seen before.

Our love of interiors can be credited to our love for architecture and design, and to our working background in product design. The way we see a product (or interiors, in this case) is a reflection of its intention, materials, making, end-use, and user. A good product closes the loop on that process. When a product is created with a true purpose in mind, it has soul; its vision has been carried through with consistency in concept, design, and making. The dots connect and a great experience is born.

In 2003 we happened upon Heath Ceramics, a fifty-year-old pottery studio in Sausalito, California, with significant history in making dinnerware and tile. The contrast between the stark design studios we had known (think industrial design and engineering) and the factory's dusty floors and constant hum of machinery excited us. Though we recognized the idea was unusual and unexpected, we were looking for something else, and it seemed obvious (to us, at least) that two designers should run a manufacturing facility. After all, good designers must intimately understand the process of making. Buying Heath led to a new way of looking at design, product, interiors, and architecture. We've always thought of Heath as "whole design"—it doesn't end with the product and involves how it's made, the environments in which it's sold, the person using it, and the buildings in which it's used. Having an understanding and appreciation for architecture, working in interiors, designing our showrooms and our factories, in turn, has helped us make better products. We can finally close the loop.

Tile is not a new idea—it's been in use for nearly five thousand years—though today feels to us relatively untapped, not typically celebrated, perhaps even unappreciated, which makes it so easy to love and so exciting to work with. There are few materials or elements in your home today that have been in use for anything close to five thousand years - maybe the stones that make up your fireplace, but certainly not the sink. The first tiles were roof tiles—simple, flat, oblong shapes of plain baked clay. The word comes from the Roman word tegula, which means "roof tile". Tile is the most durable of building materials—with steadfast colors and the ability to withstand the elements. Clay artifacts provide such a wealth of archeological information about ancient cultures because they last. But that's not the only reason. Clay and objects made of clay (tile) became a wonderful vehicle for artistic expression, especially in their use as a functional building material. We still see it in the same way as those who first explored the medium. All the qualities that made tile a long-standing building material are still valid today, hence this book.

The interiors we love have a sense of timelessness and use tile and other materials in creative ways rather than being of any particular style. We've been in spaces where tile is used as a counterpoint to the elements it surrounds, though used very quietly, it may be almost the last thing you notice. Conversely, we've been in spaces where the tile demands to be noticed first and takes on the role of a painting that covers a full wall or a floor. We've been in spaces where traditional tile is used to surprisingly modern effect. There's a selection of all these kinds of interiors, in different aesthetics, throughout this book. We've tried to share why we chose them and what it is that makes them work for us. And we know that because they use tile, they'll be enjoyed for many, many years.

TILE MAKES THE ROOM

The following spaces inspired us for many reasons. They showcase a multitude of tile and architectural styles, but in all of them, tile makes the room.

Farmshop Restaurant

LARKSPUR, CALIFORNIA
COMMUNE DESIGN

Design is material. Commune's execution of Farmshop Restaurant embodies this ideal in its truest sense, with walnut wood, copper green tiles, teal velvet, natural concrete, and blackened steel making up the bar and lounge. Each material is pure, honest, and of equal importance as you experience the space. The expanse of green tile in its classic rectangular shape and solid color is not the focal point, but a backdrop to balance the materials surrounding it and an expression of the beautiful variation of hue that comes along with hand-glazed tile. In the kitchen and surrounding the wood-fired grill, the long green wall transitions to a sea of matte white. Floor-to-ceiling and wall-to-wall white tiles define the kitchen as well as the side of the counters in the serving area. The tile work identifies the cooking areas as being as important as the dining and bar areas and gives everything in the space a sense of timeless permanence.

Emery House

BRUSSELS, BELGIUM
AGNES EMERY

This home near the center of Brussels is the urban sanctuary of designer Agnes Emery, whose company specializes in interior decorations such as paints and tiles, including the ones shown in these photos. Color is a major part of her design vocabulary. She's not fond of urban environments, so she uses an enveloping palette of blues and greens to create a soothing space that represents the nature not found in the city. The tile installations in this old house are a great example of how old and new become old but modern. The tile itself is a Moroccan terracotta that is hand-cut with a bevel edge to allow for a very fine joint, despite the irregularity of the material. This creates a tight and clean installation that, without the break of a grout line, shimmers ever more subtly with the variation in the glaze colors and surfaces of the tile. The proportions of the rooms and doorways are very tall, and likewise the tile extends all the way to the ceiling to meet the traditional molding painted in a similar palette, wrapping the room in material and color.

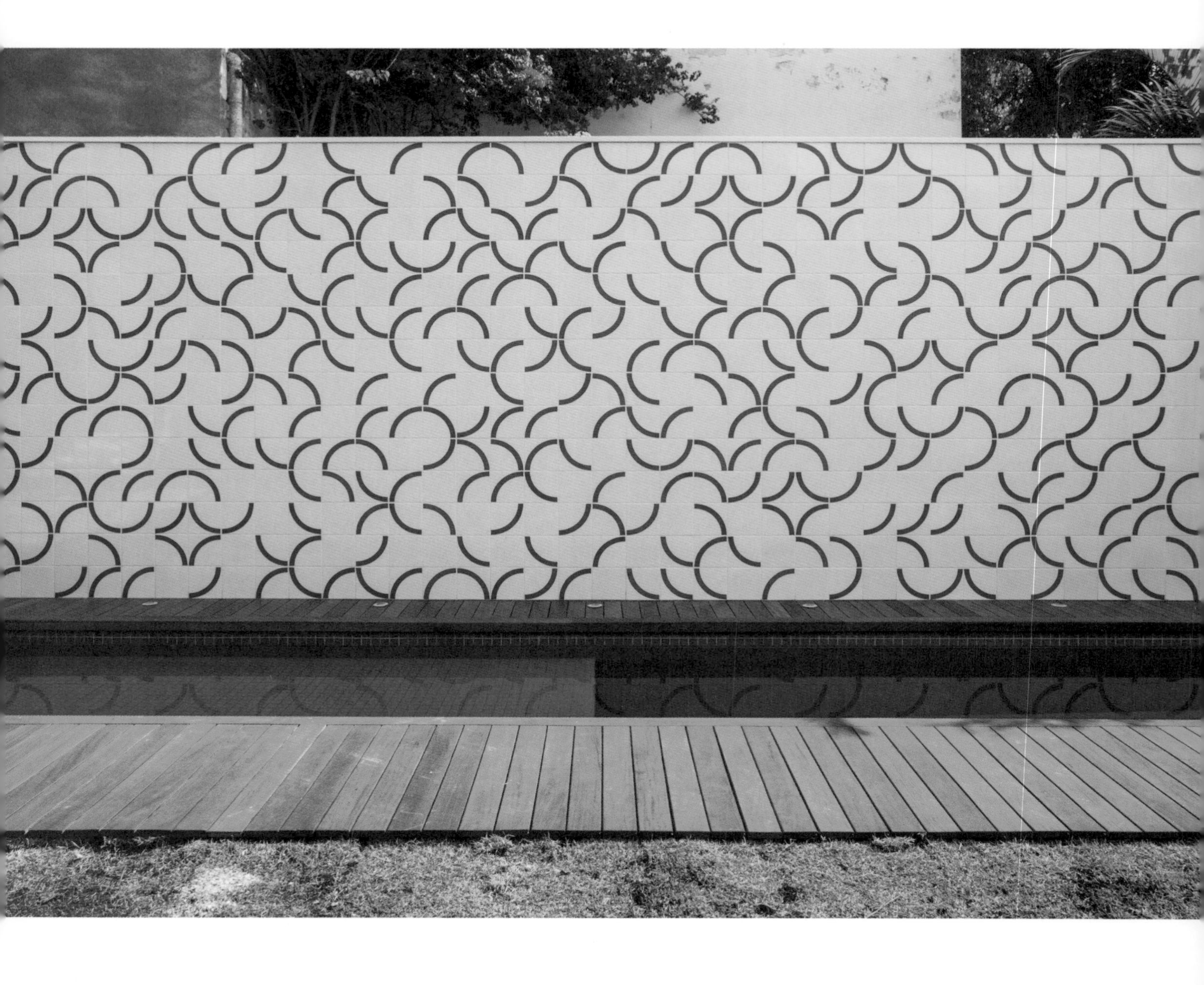

Fluid House

SÃO PAULO, BRAZIL
CR2 ARQUITETURA

Two core design elements of this São Paulo retreat are its narrow lot and adjoining garden. One of the two walls, running the length of a garden, is completely covered in graphic tiles designed by the Athos Bulcão Foundation. Athos Bulcão was an artist and sculptor, well known in Brazil for creating many tiled works that integrated art into the modernist architecture of the city of Brasília. Using a single, two-color square tile, this Bulcão design creates a mesmerising, lively pattern that not only accents the architecture, but is perceived as part of the architecture. This wall gives the exterior prominence and connects it to the social areas on the ground floor, while adding texture, personality, and fluidity that balance nicely with the glass and steel structure.

Nichols Canyon House

LOS ANGELES, CALIFORNIA
COMMUNE DESIGN

This midcentury Los Angeles home, set deep in Nichols Canyon, wholly embraces indoor-outdoor living. Commune Design reworked the entire home, in keeping with the original concept and much of its original floor plan. Tile is integrated exceptionally well in all the traditional areas, from bathroom to kitchen, even expanding outdoors to pull together the property. It consistently delights at every turn. The bathroom on the right has floor-to-ceiling 2x4" tiles in a slightly metallic, onyx glaze. The room comes alive through its combination of materials, and is grounded by dark tile that plays with the light pouring in from a glass ceiling. The designers used the same size tile in the outdoor bar and barbecue area shown above, this time in a dark brown that works with the natural wood decking. The focal point of the exterior is a massive wall fountain, by ceramic artist Stan Bitters, that connects to the swimming pool (see page 60). The vertical sculpture was made specifically for the site by Bitters, using his iconic three-dimensional tiles. On the following spreads (pages 60–63) are details of another Bitters ceramic tile wall sculpture and the fully tiled master bath. The success of this room's design comes from combining an interesting pattern in the tile layout, a gray and white palette, and the tiles that cover all the surfaces of the room. It's neutral and understated, while interesting and strong in it's reference to the house's roots. One other tile wall fountain (not pictured) was commissioned for an indoor/outdoor dining area on the other side of the house, creating delightful visual consistency.

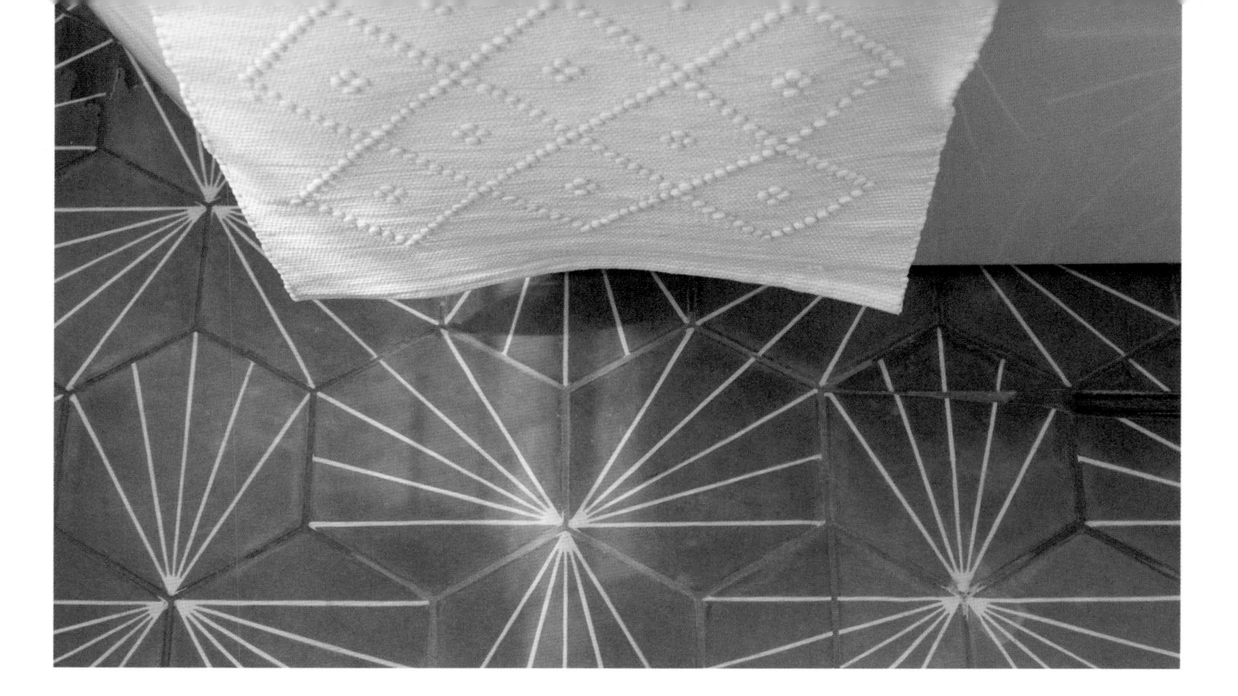

Murnane House

LOS ANGELES, CALIFORNIA
PROJECT M+

This inviting bath, in the Los Angeles home of creative couple and principals of Project M+, McShane and Cleo Murnane (he's an architect, she's a designer), is located on a hill with expansive views. The bath, however, is a windowless room with an oversized skylight, strategically placed to flood the space with light, bringing the outside in. The wall tiles create a simple, bright white backdrop for the room's main design element: the floors tiled in a modern Moroccan cement tile with a graphic pattern. The result is a contemporary nod to a classic shape and material. The blue tiles ground the space and all elements work together to keep a bright and airy environment.

Cristalli Bath

COPENHAGEN, DENMARK
MADE A MANO WITH DORTE HØEGH

This bath sits on the second floor of a small three-story house. Since the family's four children have their own bath in the same tile (theirs in white), the homeowner went all out on color with hers. The rationale was that small spaces need something special. The herringbone patchwork dominates the space, while the scale of the tile is large enough to feel playful and expressive. The bright, shiny glazes contrast with the tile itself, a lava stone from Sicily. While the lava stone is a somewhat coarse natural material, it's been cut to precise rectangles with a smooth flat surface. The colorful glazes paired with coarse, natural materials is striking in richness and warmth (important in a bathroom, especially a small one). The essence of lava stone remains under the glazes so that it pulls together all the colors into one family.

Studioilse

LONDON, ENGLAND
STUDIOILSE

In the headquarters of Ilse Crawford's London design studio is a small kitchen and eating area. Housed in a nineteenth century building, this location presents the type of design and space that we really like: functional and understated, with beautiful detail and an easy relationship between old and new. Crawford's mission to put human needs and desires at the center of her interiors resonates with us. In the design for the kitchen in her studio, she let the age and character of the building be seen in exposed pipes and brickwork, while a few beautiful design elements are inserted to show her studio's aesthetic and point of view. The classic style of the gold pendant lamps, combined with furniture of her own design, make this tiny area a functional one that's rich in character. The island and kitchen area are clad in simple white tile, while the area with the dining table is white painted brick. The honesty and functionality of the tile are just right for this design, and by using a single material to clad all the walls and island in the kitchen, an easy distinction between the areas for cooking and eating is created.

HOT
COLD

Catalina House

LOS ANGELES, CALIFORNIA
COMMUNE DESIGN

The family home of Ramin Shamshiri and Donna Langley is a 1920s Spanish-style property in the hillside Los Feliz neighborhood of Los Angeles that overlooks the city. The home had lost its way after several remodels by previous owners. Shamshiri partnered with his sister Pam, also of Commune Design, to reimagine and restore the house's strong Mediterranean bones, infusing it with their Californian point of view. The courtyard's fountain and Moorish-inspired tile pattern connects to the kitchen. The black and white concrete floor tiles, rich green cabinetry, and walnut counters make up the home's soul, as the patterned tile flows from the kitchen to the outdoor eating and lounging area on the courtyard. Tile is used as a decorative element that emphasizes the original architecture of the home, and also to connect different spaces. A delightful hanging curtain of clay tiles forms a screen to distract from the property next door. The curtain hangs free, acting as wind a chime when breezy, and adding a wonderful element of sound to the space.

Hotel Okura

TOKYO, JAPAN
YOSHIRO TANIGUCHI, HIDEO KOSAKA, HAJIME SHIMIZU, AKIRA IWAMA, AND KISABURO ITO

Built in 1962, this icon of Japanese modernist design may well be torn down by the time you read this. It was built just two years before the first Tokyo Olympics and is now making way for a larger hotel for the 2020 Tokyo Olympics. For this reason, we felt we had to include it in the book and jumped through some last-minute hoops to get the photographs. For devotees of the hotel, the design exemplifies a particular moment in Japanese culture, which thankfully has remained intact through its five decades of operation. While the design is modern, there are also elements of traditional Japanese crafts and materials, by well-known craftsmen of the time, including these tiled walls in the public stairwell. The tiles are traditional in design, but feel modern in the space. The wall spans the entire stairwell, not simply stopping at the first landing. This defines the space in a committed way. The variation in color, the irregularity of the hand-cut shape, and wonderful pattern created by the varied bisection of the rectangular tiles gives a softness and strong sense of materiality to this public space.

Stand-Alone Bath

PARMA, ITALY
FRANCESCO DI GREGORIO AND KARIN MATZ

Located on a tiny street in the middle of the small town of Parma, Italy, Francesco Di Gregorio and Karin Matz converted a stable into an apartment for a young couple. Existing elements, windows, walls, and columns remained unchanged while a bathroom was added that was entirely freestanding within the original space. The bathroom "box" is completely covered in 4x4" tiles, creating a structure in which space has been turned inside out. The glossy tiles bring bright light and uniformity to this exciting clash of old and new architecture. The solidity and contrasting material of this new element affirms its identity as an addition to the space, rather than trying to integrate with the construction of another era.

Wedel House

SACRAMENTO, CALIFORNIA
POPP LITTRELL ARCHITECTURE + INTERIORS

This is one of those happy accidents that occur when a designer is given a constraint rather than a clean slate from which to design. You'd never guess that designer Curtis Popp's parameter was his client's red Bertazzoni range, bought on clearance, just before he dug into the design. Since the color palette was set, the kitchen design needed to evolve from the range. Next up, he found a hand-glazed red hexagon tile with nice variation in color from Heath. As luck would have it, the tile's color connected directly with the red of the range. Stained wood paneling and shelf brackets made from branches create a nicely balanced design with a striking red center point, a great example of small-scale color creating a big impact.

Heath Ceramics Los Angeles

LOS ANGELES, CALIFORNIA
COMMUNE DESIGN AND CATHERINE BAILEY

The Heath L.A. showroom was not designed to sell tile, yet tile was important to the space's design. Commune developed a tight material and color palette and we agreed that the tile must fit into the palette to keep the design vision focused. The showroom was built on honest materials, like knotty pine, ceramic tiles, and concrete, and a color palette of Heath's signature orange, blues, and grays. Panels of dual-glazed orange tile line the front of a counter, making tile part of the furnishing along with the architecture and floor-to-ceiling tiles of four different blue glazes lining the 15-foot high wall in the back of the space. Another goal was to share the story behind the products. A portion of the Los Angeles showroom is also a clay studio, where those pots on the tiled counter were hand thrown.

Heath Ceramics San Francisco

SAN FRANCISCO, CALIFORNIA
COMMUNE DESIGN, CATHERINE BAILEY, AND THE HEATH DESIGN TEAM

The Heath San Francisco location has many spaces, which were designed and executed over a three-year period. The original palette of material and color was developed and then referenced when moving into the next area. This approach provided consistency, along with the freedom to explore the needs and wants of the space. Tung Chiang, Heath's studio director, created a unique design for his studio's work area. The pattern, shown above, of little diamond tiles in different glazes is repeated every 8 feet, giving the impression that the design is more of a custom mural than a repeatable pattern. The palette in Chiang's studio is narrower than the one used in the showroom—the palette of whites and grays with light natural wood makes his space feel unique, yet still part of the bigger Heath aesthetic that runs throughout the sixty-thousand-square-foot building. Commune designed the showroom and its stunning kitchen, and both are heavily focused on tile. Four blues, matte and glossy, are used on the large walls that frame the kitchen and extend all the way to the tall ceilings, creating a surprising and delightful interplay of light that reflects and dances around the room. A "crease" three-dimensional tile panel is integrated into the center island. Framed in wood, this tile panel can be swapped out, and removed the need for potentially fussy details at the edges. Pages 88 and 89 show our employee kitchen, which has an experimental screen-printed tile that we felt a need to install on a large scale in order to understand the full potential of the design. Featured on page 89 is a detail of the tile wall behind the coffee kiosk at the building's atrium, made up of a palette of mixed shapes and colors that tie into the family of glazes used throughout the showroom.

BEGINNINGS

6
AMERICAN RANGE

Beachwood Café

LOS ANGELES, CALIFORNIA
BESTOR ARCHITECTURE

The first time we walked into the Beachwood Café, the interior felt visually surprising, in an odd but good way that we don't experience as often as we'd like. It's an unpredictable combination of elements, all with their own personality, that somehow works. Architect Barbara Bestor renovated Beachwood Canyon's beloved Village Coffee Shop, salvaging the best of the old rustic interior, updating some details, and adding personality. It's another good example of how spaces can evolve and move forward without the need to delineate between old and new. The vibrant, geometrically patterned yellow and blue cement tile on the floor demands your attention and shifts the color palette up the wall to the painted wainscoting, before the Geoff McFetridge patterned wallpaper takes over to tell a story of its own. Paired with some original knotty pine details and vintage tables, the design is fun, cozy, stylish, yet unpretentious and not at all typical or trendy. It feels like a cool friendly neighborhood spot that will stand the test of time.

Popp House

SACRAMENTO, CALIFORNIA

POPP LITTRELL ARCHITECTURE + INTERIORS

Located in sunny Sacramento, this colorful bathroom is in the home of designer Curtis Popp. He chose this house for its early 1940s art moderne architecture and has gently updated some of the areas so that it still reads 1940s while also reflecting his taste for modern design and vibrant color palettes. The interior and exterior of the house are painted white, creating a clean, neutral backdrop for vibrant spots of color and design. Popp has successfully explored these ideas by creating a striking focal point in this bathroom with Heath hex tile in a rich "Tropics Blue" glaze that is strikingly colorful while also referencing water and the soothing nature of bathing. The translucent crackle glaze has depth and personality from the variation in the hand glazing of the tile, nicely merging with the patina of the house's 1940s details.

Upstairs Office and Apartment

Heath San Francisco

SAN FRANCISCO, CALIFORNIA
CATHERINE BAILEY AND THE HEATH DESIGN TEAM

We're not sure what to call this space, as it's a catch all for many aspects of our lives. Located in the same building as our San Francisco factory, it serves mostly as an office and hideaway for the two of us, perfect for when we need to get away and focus or just clear our heads. But as a couple that works together, lines between home and work become very blurred. Sometimes when we're working here, our son is playing with Legos, and the dogs are sprawled on the floor and sofa. The design is formal enough to have a meeting, while the absence of a single item of office furniture made it comfortable enough to work on this book while being sprawled out on the couch (appropriately, it's also where we keep most of our books on design). The tile on the kitchen wall extends to the ceiling and defines that end of the room, with the shelves and flue sitting atop the tile as secondary elements. The tile, with its volcanic texture, was chosen to soften the space's concrete walls and to work with the original rough plank floors, so that there's a comfortable progression between the old and new elements. The bright reds and blues of the tile wall mural provide a colorful counterpoint to break up the concrete and raw wood hues of the materials that define the space. Every element is in keeping with the overall building's evolution; new materials working with old materials that have stood the test of time.

AMERICAN RANGE

katachi
JAPANESE PATTERN AND DESIGN IN WOOD, PAPER, AND CLAY
WHARTON ESHERICK
Design Research
WHY DESIGN NOW?
Pratt
China and Glass

Alabama Studio Style
Planning & Landscaping HILLSIDE HOMES
TRUCK nest
A RECORD: NINE YEARS IN THE MAKING
A Pattern Language

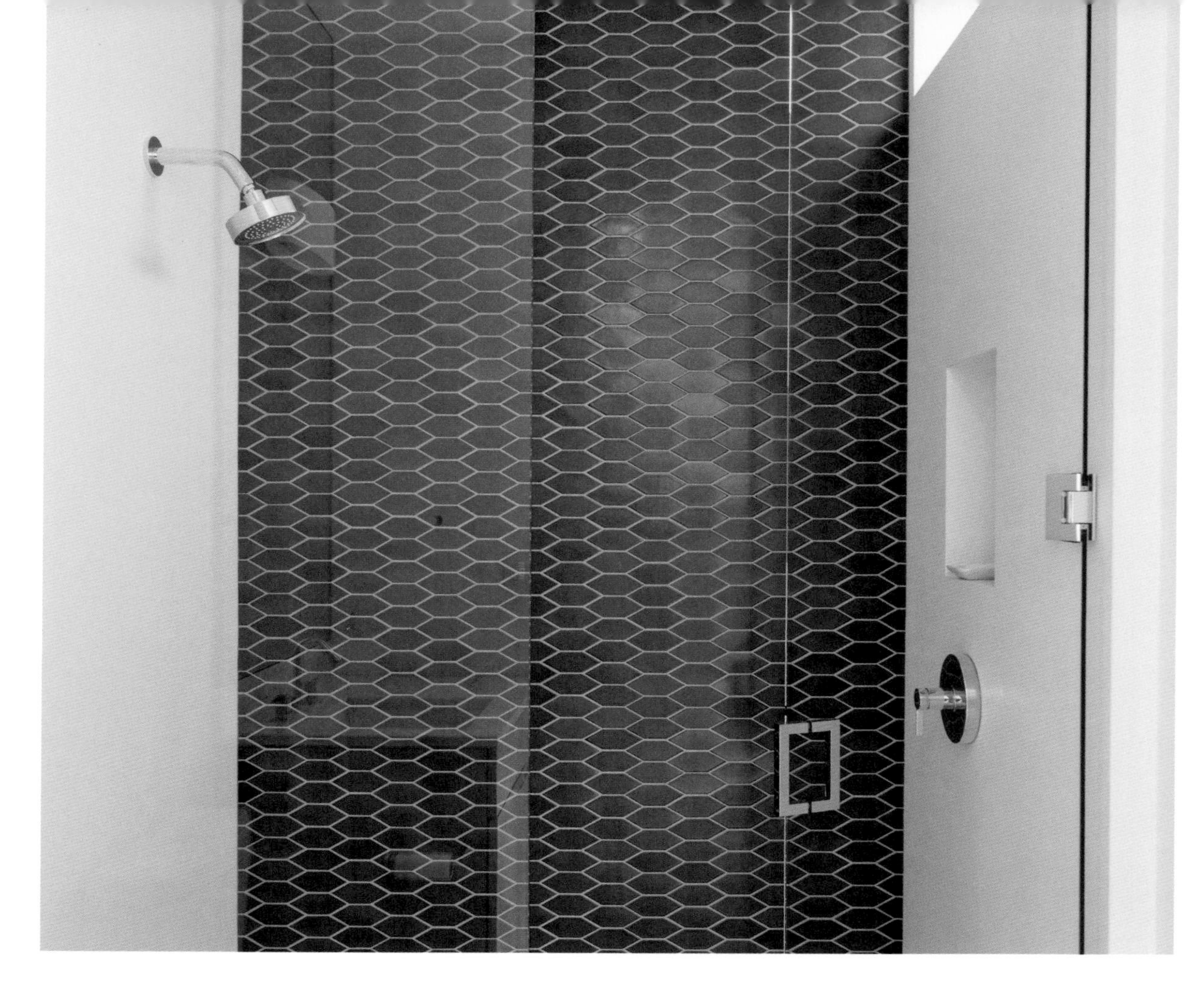

Butterfly House

SAN FRANCISCO, CALIFORNIA
JOHN MANISCALCO ARCHITECTURE AND SHAWBACK DESIGN

The Butterfly House is located on a prime lot in San Francisco with a view of the Golden Gate Bridge and beyond. Though it is a rebuild of a midcentury home, Maniscalco created a structure that's rooted in the future. The overall design is modern, airy, and pristine, with a bright white and neutral color palette. The views are its focal point. Though the design is modern and very precise, handcrafted tiles in different shapes and neutral colors create warmth and successfully add diversity and texture to the overall design. Notice the details, like the tile that extends behind the open cabinets in the all-white kitchen to create complete layers of material. Each shower in the house uses pattern to provide and define character. Custom concrete panels cover the adjoining surfaces, creating seamless joints, which allow the tile pattern and grout lines to become an even stronger design element.

KitchenAid

Green Light Kitchen

MARINA DEL REY, CALIFORNIA
DISC INTERIORS

We like the simplicity and restraint that Disc Interiors used here. The right materials in the right proportion make this room stunning. A simple green tile is installed in a subway pattern cladding the entire kitchen wall and the island. This creates a simple and singular visual plane of tile as you view the space from the dining area. The choice of glossy tile to reflect the light from the adjacent wall of windows was a decision that adds to the interesting character of the space without adding complexity.

Parco dei Principi Hotel

SORRENTO, ITALY
GIÒ PONTI

We could go on and on about Italian architect Giò Ponti's masterpiece in the Parco dei Principi Hotel. This space expresses so much of what we love about materials and design creating a sense of place. This iconic hotel, designed in 1962, is one of the finest examples of an all-encompassing Ponti aesthetic. The large white modern rectangular building is anchored to a cliff with majestic views of the sea. Ponti was involved in every detail of this hotel, from the furniture to the lighting to the thirty different geometric-patterned tiles. The palette for the entire hotel is blue and white drawing from the sky, sea, and islands. Ponti designed the thirty 20cm x 20cm blue and white tile designs to be arranged in different combinations to give each room its own story, drawing on patterns from moons to leaves. All the flat tiles are the same size majolica tile, hand-painted and manufactured locally by D'Agostine, a factory in Salerno. What makes the overall design even more interesting is that Ponti pushed the idea of tile beyond the common form of flat tiles. On the walls of the reception and dining area are thousands of shiny blue and white glazed ceramic pebbles embedded in a matte white mortar, creating an organic texture that looks as if it may have grown out of the sea. In these same spaces, other walls and columns are covered in ceramic tile plaques of various forms and glazes created by Italian figurative sculptor Fausto Melotti, who also worked with Ponti on several large projects. The shiny tiles reflect light and create texture on otherwise flat surfaces, resulting in modernist tile artwork that blurs the lines between art, interiors, and architecture.

Kogure House

TOKYO, JAPAN
JIRO MUROFUSHI

This home feels very Japanese in scale and material, until you look at the tile. The homeowner, a photographer, found this handmade tile on a trip to Fez, Morocco, and brought it home to Japan. The result is a space that may not at first feel aesthetically Japanese, but honors that culture's appreciation for good materials that acquire a natural patina over time. The irregular shapes of the tiles create the casual environment that the owner was seeking—he describes himself as someone who prefers a bicycle over a Rolls-Royce. The thick white tiles on the kitchen counter and walls have a beautiful worn look that set the tone for a well-used kitchen and its well-used implements. On the floor, the herringbone pattern in blue and white has an equally beautiful patina, with a wonderful border at the edge of the counter that creates the frame a herringbone layout needs. Blues and whites were chosen for the floors to give the feeling of a rippling water channel through the narrow hallways and spaces that meanders the pathways of the home. The irregular mixing of blue glazed tiles in the pattern also create a feeling that's casual in a house that's clearly more a home than a showcase.

MAD House

VANCOUVER, CANADA
MARIANNE AMODIO ARCHITECTURE STUDIO

MAD stands for multi adult dwelling, as this home is lived in by three sets of adults from the same family. The first thing to note is the bright yellow door, window frame, post, and mailbox in contrast to the boxy white facade of the house. The glow of yellow becomes more intense the closer one gets to the front door, and looking up, one is delighted to find the walls and ceiling of the foyer tiled bright yellow, interspersed with individual tiles in green (for summer) and red (for fall) in a fish scale pattern. The entry foyer cutout only strengthens the feeling of standing inside a jewel box. A very different, but equally gorgeous tile installation is found around the fireplace. In a spare interior of concrete floors and white walls, the clean rectangular fireplace is clad in a geometric pattern of tiles in bright glazes, ornate shapes, and hand-painted traditional graphics. The overall effect is reminiscent of traditional Mexican tile, but installed with a modern sensibility.

CHRISTOPHER WOOL
WILLEM DE KOONING: PAINTINGS
robert Rauschenberg combines

Stark House

SAN FRANCISCO, CALIFORNIA
JONES | HAYDU

The renovation of Janna Stark's beloved San Francisco home came after it suffered extensive fire damage. Before the renovation, the home had lovely original Victorian details, but an awkward flow. The renovation presented the opportunity to open up the living spaces. Jones | Haydu relocated the main seating area to the street side and moved the bedrooms to the back. But what really set this design apart from other modern spaces was its rich, natural materials—they salvaged charred wood from the fire and reused it, installed Douglas fir cabinetry, and made the focal point a fireplace clad in three-dimensional oval tiles from Heath. The proportions and location of the fireplace are in keeping with traditional Victorian flats, yet the midcentury style of the tile and warm hues tie it to Stark's love of midcentury and Danish Modern furnishings. The fireplace's clean rectangle allows the tile to extend from the firebox to the ceiling, and its wooden end panels and a stone surround remove the need for fussy details, allowing the tile to live on the wall as art.

White Brick House

PORTLAND, OREGON
JESSICA HELGERSON INTERIOR DESIGN

This project, a 1920s Mediterranean-style home built of white brick, opened up the kitchen to create a more spacious feel and connect it with the rest of the home. You may not first notice that every bit of the wall, including the archway between the dining room and kitchen, is tiled. It's a subtle use of material, because being in the space one may feel the materiality before seeing the tile. The tile on these surfaces is a pretty simple glossy white. It wouldn't feel out of place in a classic kitchen, yet it introduces an element of surprise because of how extensively and consistently it's used throughout the space. This commitment to the material also helps maintain the strong identity of the kitchen, despite the open floor plan. The surfaces where one is used to seeing tile, like behind a sink in a kitchen, blend right into places where it's less commonly seen, like all the way up to the ceiling and above and around the window. The surface of the tile is not perfectly flat so one feels the materiality, and this is a huge part of the feeling of the space. The graphic tile on the floor lends itself equally to a classic kitchen feel, while serving as a foil for the white tile on the walls. It catches your attention first and lets the subtle tonality of the tiled walls surprise and delight you even more.

Waters House

BERKELEY, CALIFORNIA
ALHORN/HOOVEN

Go into Alice Waters's famed restaurant, Chez Panisse, and you'll find that every detail, from the food to the setting, is carefully considered, rooted in materials, and works together in harmony. It's no surprise one finds the same in her home, a modest Craftsman in Berkeley, California. The green tiles wrap around the corner of the kitchen where the stove sits, and set a natural organic color palette that's at home with the Craftsman style. The earthy green tiles frame a corner where copper teapots sit atop a well-used functional beast of a gas range. Copper pots and trays filled with cooking implements collected from travels over many years sit beside it. A concrete counter completes the theme of materiality, earthiness, and texture that come only with good materials that have been well used. Moving along to another part of the kitchen, smaller tiles mixed in a similarly earthy palette frame a kitchen window that opens to the garden outside.

Hillside House

MILL VALLEY, CALIFORNIA
SB ARCHITECTS AND ERIN MARTIN DESIGN

This vertical home, built by architect Scott Lee for his family, is carved into a steep Mill Valley hillside. Its four floors make up a modest 2,100 square feet, yet the the exterior space and connection to the views create a comfortable and dramatic environment. Each floor takes in the view from the valley to the bay. The bathroom shown here uses a simple, plain white glossy tile to frame the view, the different size tiles in the niche create a thoughtful detail, and the overall material combination with dark window frames and concrete surfaces pull together a rich look through the use of neutrals. A bath at the other end of the home combines darker shades of neutrals and uses a double-glazed Heath tile to keep the surfaces interesting. The glossy and matte glazes combined on individual tiles breaks up the traditional grid and create an engaging surface that comes alive as the light hits it.

McKenzie House

BORREGO SPRINGS, CALIFORNIA
MAURICE MCKENZIE AND STACEY CHAPMAN PATON

The complementary diamond and bow tie tiles used in this home's kitchen were originally found tucked away in the slip casting area of Heath's 1959 factory. When we brought them out of retirement, this installation was one of the first to feature these three-dimensional tiles. In this home by architect Maurice McKenzie, the owner updated the kitchen in classic midcentury style. The large tiled rectangle above the kitchen counter is a major aesthetic element on its own. In an uncluttered layout where the kitchen opens to the dining and living areas, and provides generous sightlines from various vantage points, it is an example of materials being appreciated for what they are, beyond the functionality they provide. The frost glaze used is a period classic as well, with high variation and translucency that lets the color and texture of the brown clay show through. Combined with the concave surface of the diamond tiles, the light creates a wonderful, lively visual interest, in which no two corners of the wall are the same. On the kitchen counter, rest objects with a midcentury pedigree, but because none of the objects are tall, the tile stand out as more of an object than a backdrop.

Komon Bath

HELSINGØR, DENMARK
MADE A MANO

The tiles that make the foundation of this airy and bright bathroom are designed and made by Made a Mano. They add visual weight, grounding the minimal space. The tiles are made from Sicilian lava stone, cut to size, while the pattern is a glaze that's been screen printed, then fired. The combination of the natural lava stone with a traditional Japanese small decorative pattern gives a sublimely organic but refined aesthetic to this bathroom floor. The scale and variation of the detail within the sparse and minimal space pulls together the design and adds the warmth of hand-made detail that's needed to keep it balanced and interesting. Another nice detail is the floor line that transitions into a recessed tub and shower, providing a strong graphic presence that makes an impact on the space.

Zamora Loft

OAKLAND, CALIFORNIA
CHRISTINA ZAMORA

Created by Christina Zamora for her home in Oakland, this bath evolved from the tile itself. While a designer at Heath, Zamora was involved in designing the colors and patterns for this line of tile. When designing, it's always helpful to have context to design around, and so her own bathroom became the perfect blank slate to imagine the new tile patterns. Cool grays tie into the modern, minimal aesthetic of Christina's loft, balancing a bright yellow floor pattern as the focal point. The gray transitions the focus into concrete walls, completing a resolved, clean, wall-to-wall pattern.

Underwater Bath

SAN FRANCISCO, CALIFORNIA
ENVELOPE A+D

The homeowners are skateboarders with an irreverent design sensibility. In their home's bathroom, they wanted to re-create the sensation of dropping into an empty pool and skating in its bowl-like volume. Tile helped achieve this effect, with its solid feel and choice of colors speaking the language of pools and water. The floor and wall are joined seamlessly by the tiled pattern, forming a rounded L shape. Envelope A+D used a Bisazza glass tile with an embedded fade in the pattern. Envelope tweaked and customized the original pattern so that it and the gradient extended, and solid colors were placed at either end of the spectrum. The distinct chromatic shift from ceiling to floor was integral in achieving the effect desired by the design concept, creating a sense of increasing depth as one moves down toward the denser pattern of blue tiles. The tile marks a depth of ten feet near the ceiling to solidly assert the intended pool context. Where a "pool in the bathroom" concept could come off as cliché, this example shows that use of appropriate materials, thoughtful details, and excellent craftsmanship can make a design both unique and timeless.

Yardhouse

LONDON, ENGLAND
ASSEMBLE STUDIO

The architects designed this building as an affordable creative workspace, featuring individual studios and communal spaces. A simple structure that's relatively cheap to build and assemble (and even disassemble and move elsewhere) creates the affordability. In keeping with the spirit and vision of the use, not for quiet desk work but for a range of messy and noisy tasks, the concrete tiles used to shingle the exterior facade were all handmade on-site. The beautiful range of colors was created by adding pigment to the concrete, but its the process that stands out. No measuring, no repeating, no standardization—whoever was making tile at the time just added pigment to each batch, however much they liked. The same loosely creative process was followed in hanging them. The result is a beautiful, randomly colorful wall of art pulled together by the consistency in shape and tonality of the concrete. The choice of materials and the creative spirit in which they were made work so well with the intent of the building and how it's used.

Llama Restaurant

COPENHAGEN, DENMARK

BIG, KILO AND HZ

Designing the restaurant Llama presented the challenge of merging South American and Danish cultures. Jakob Lange of design studio BIG describes the result as, "a remarkable transformation that has turned traditional Latin American vernacular into a contemporary Danish public space." The tile is an important element in its success, as it is a handcrafted Mexican tile that fuses Latin and Scandinavian designs. Lining the floors and walls of the three dining spaces, the tile creates an intimate and connected space that feels distinctive yet integrated.

Roddick House

LONDON, ENGLAND
MARIA SPEAKE, RETROUVIUS

While the tile used in this home feels classic and old, the space feels modern because of how the materials are used. It's clearly well lived in by its owners. The full wall of tile behind the kitchen counter and appliances lets these elements take a "smaller" role in the room without having to build them in with cabinetry to hide them. The textures build on one another, such as the brown tiled fireplace, with the fabric fringe finishing off the base of the stone hearth, continuing that modern feel through unexpected combinations of materials. The chalkboard in the entry hallway signifies that ever-changing lives and schedules happen here. The black and white patterned tile wrapping around the floor and to the base of the chalkboard is an important choice—it's a busy pattern that makes the natural clutter of life in this room work in a way that a more pure background would not.

Pluijm House

MARRAKECH, MOROCCO
ANK VAN DER PLUIJM

This home's design is driven by the materials used. All of the choices have been made very carefully, and the tile gives the space a feeling of solidity that other materials wouldn't, while adding a decorative element. The mix of green patterns on the side of the kitchen counter create a lighter visual interest where there would otherwise be a solid mass that wouldn't balance out the solid column to the right. In all the spaces, particularly behind the kitchen sink, the tile has been fitted together pretty tightly so that grout lines are not prevalent, allowing the variation in surface texture and color to create the depth and pattern in the installations. This is especially important given that natural light is a major design element in all the rooms, and the tiles chosen, with their inherently soft, handmade feel, are only enhanced by its presence. The looseness in irregularity of shape is fitting with the rest of the spaces, from the wooden reed ceiling to the plaster walls. Greens are a common hue, and in the bathroom, the subtle pattern of different shades creates a charming band across the room.

Zittel House

JOSHUA TREE, CALIFORNIA
ANDREA ZITTEL

Andrea Zittel is an artist living in the high desert of Joshua Tree, California. In 2000, she established A-Z West, a testing ground for experimental designs for living. Aesthetic sensibility and "investigative living" come together in her home where, she incorporates tiles with a strong graphic pattern that's repeated in other ways throughout the living spaces. Zittel originally created the tile patterns as gouache paintings on paper—she was interested in the repetition of patterns and their ability to create infinite surfaces. Inspired by early modern movements like the Russian Constructivists and De Stijl, the tiles also echo explorations into pattern and brickwork by Josef Albers. Zittel says that she is particularly fond of the pattern's ability to camouflage a slightly dirty floor.

FINE Design Office

PORTLAND, OREGON
BOORA ARCHITECTS

This striking kitchen is housed in the creative offices of FINE Design Group in Portland, Oregon. Boora Architects created a loft-like open environment with a relaxed feel. This office kitchen is framed by a fully tiled wall and floor that are made up of two colors in two different widths. The result is a surface that's refreshing and unique. The installation is also an example of what we mean when we say Heath tile is perfectly imperfect. Each tile is a slightly different length with slight bends, yet the overall surface and design come together beautifully—it's warmer and more organic than tile that was dry pressed by a machine.

Peaks View House

WILSON, WYOMING
CARNEY LOGAN BURKE ARCHITECTS

This Wyoming home is surrounded by nature, and the materials used fit in quite well. The owners chose the glaze colors after their pilgrimage to Heath Ceramics. The outdoor cooking area is tiled in a brick color and paired with unfinished wood siding. It uses a traditional brick layout for continuity with the exterior house wall and present a modern take on a cooking hearth. The theme of bold color with high variation is continued into the shower and bath, this time in an aqua blue. The larger-scale tile, closer to a brick's proportions, create a solid alcove that feels far from delicate but still fits into the construction of the room. The floor-to-ceiling tile wraps around, enclosing the space, giving it a surprising warmth and coziness. The alcoves in the shower are planned to accommodate the dimensions of the tile (a detail we highly recommend, when possible), keeping the visuals across the wall clean.

Margarido House

OAKLAND, CALIFORNIA
ONION FLATS AND MEDIUM PLENTY

This home was the first LEED-H Platinum certified custom home in Northern California. Located in the Oakland Hills across the bay from Heath's factory, it was the first project that used our kiln shelves (furniture that once held Heath tiles as they fired in our kilns). The shelves had come to the end of their useful life and were glazed and repurposed into exterior wall cladding and a patio floor. Several of the bathrooms in the house also used Heath tile, stacked in a modern grid pattern, with strong color and high variation to soften the crisp spaces. The use of floor-to-ceiling tile is also important in keeping with the modern practice of using large, single planes of material. Even the niches of tile in the showers are consistent with the design. The mix of natural light, colorful glazes, and water create relaxing spaces for bathing.

São Paulo House

SÃO PAULO, BRAZIL
GUILHERME TORRES

This 1970s-era residence was upgraded for a family of five relocating from London, who were bringing all their furniture with them, which would serve as a reference for the design. In this home, the same tile is used to achieve decoration on the interior and to create an architectural element on the exterior. In the open-plan dining room and living space, a long wall is patterned with a graphic tile of vibrant colors that go from the floor to the ceiling. This large rectangular backdrop is uninterrupted—a modern design element covered in a decorative tile. The tile is used the same way as wallpaper or paint, but the material choice makes this wall a solid anchoring element. The same tile is used to clad the rectangular volume of a hallway protruding from the building, opening onto the garden and pool area, and creating a nice connection between these indoor and outdoor living spaces that the family enjoys together. As with the interior placement of the minimal concrete dining table, the decorative and vibrant tile pattern contrasts the clean simplicity of the outside concrete walls.

The Beatles

Romita Comedor

MEXICO CITY, MEXICO
RODRIGO ESPINOZA, MARCELA LUGO, AND ARTURO DIB

This restaurant is located in an old Mexico City neighborhood, in a building dating back to the early 1900s—it's currently protected as a historic structure. The materials used in this updated space—wood, glass, tile, potted plants—resonate with the building's history as materials of quality that will gain a nice patina over time. The floor tiles are original, covering what was the roof terrace when the house was first built. While we don't typically love an overt graphic like a white letter R on a black tile, the abstract pattern it creates is an interesting element and backdrop in this old-meets-new space. The consistency in the color palette and the commitment to tile, even on the doors, contributes significantly to the overall feel. It is an aesthetic that feels modern, even whimsical, but because of how well it's integrated with all of the other elements, it's likely to age well over time.

NO FUMAR

STUMPTOWN

Ace Hotel

LOS ANGELES, CALIFORNIA
COMMUNE DESIGN

This historic United Artists Building is a Spanish Gothic structure built in 1929 and renovated in 2014 as the Ace Hotel in downtown Los Angeles. Commune Design worked on the interior and concept, combining Hollywood glamour with modern minimalism. As seen here, the restaurant and foyer, blur old and new with many references to the past. In the lobby, the designers left highly detailed moldings around a doorway and the original floor, while inserting custom tiles they designed in white and black below the counter of the new coffee kiosk. Those same tiles make their way into the restaurant but with the colors reversed. In both cases, the geometric lines work with the original details, bridging the gap between new and existing elements. Lining the walls of the rooftop bar are concrete tiles with a molded geometric pattern, acting as wainscoting. These custom tiles create a layer of detail in the same material as the original concrete structure, which dominate the space, and create a more intimate environment.

Avenue Loft

PORTLAND, OREGON

JESSICA HELGERSON INTERIOR DESIGN

This home is in a converted early twentieth-century warehouse and manufacturing facility that was made into condos in the 1990s. The original conversion left a very modest 870 square-foot space that was very divided. The new owners wanted an open space where they would feel compelled to cook and entertain. Jessica Helgerson took on the project with the goal of creating a warm and sophisticated home taking into consideration the context of the old building. The choice of glazed brick rather than traditional tile ties beautifully to the original building materials and the industrial purpose for which the structure was built. In the window above you can see a bit of the original brick lining the sill. The integration of the hood into the wall by covering it with the same glazed brick makes the whole kitchen feel solid and connected to the actual architecture of the building. Additionally, the neutral color palette still allows for contrast and richness in the design, creating an inspired space in which to spend time cooking.

Bailey-Petravic House

SAUSALITO, CALIFORNIA
BARBARA BROWN AND CATHERINE BAILEY

This has been our home, project, and hobby for more than a decade. Before it became ours, it was likely lived in by at least four generations. Built as a simple retreat for a deckhand in what was rural Sausalito, an area that was close to shipyards and docks. Originally it had no plumbing or electricity. It has, of course, evolved over time, though still has its original footprint and proportions. Its proportions and character are what drew us to it, along with its views of the bay. Our first project created a lower level from an unfinished basement, and features the green tiled bathroom shown on page 20. We used second-quality tile from a large job, installing it in a running bond pattern that keeps the eye from focusing on an imperfect grid. The main floor is a sea of color and pattern, with a yellow tiled kitchen of beautifully varied glaze that is the result of a happy accident on a large commercial job. The upper floor is the most recent project, where by adding a dormer we were able to add a small bathroom and second bedroom on the same floor as our bedroom. The colors on this floor are neutrals, and mostly whites. The bathroom is entirely clad in a variety of whites in different finishes, which is meant to be a peaceful break from the bright colors and patterns found on the floors below. We wanted a quiet, yet interesting, experience for this room, and something we could live with forever. The tile in our home also extends outside. The pond you pass before you enter the house holds a running bond pattern, a nod to the brickwork on some areas of the house. Heath House Numbers have been integrated into the pond's tile design, as well.

Turner House

LARKSPUR, CALIFORNIA

JENSEN ARCHITECTS AND NICOLE HOLLIS INTERIOR DESIGN

The simple shape and texture of penny-round tile makes this bathroom a classic. Penny round tiles have a natural affinity to bathrooms, perhaps because of their tactile nature. The texture of the surfaces work well, softening the long rectilinear lines of the room, a space that, by its nature and association with bathing, needs to feel both clean and enveloping. The extension of the tile on the entire back wall of the bath keeps that element of cleanness and modernity. It's also in keeping with the long rectangles of the mirrors and vanities on the opposite wall. On the window, a nice feature of die-cut circles in an aluminum panel echo the shapes of the tile and form a privacy screen that creates modern character, as well as creating ever-changing shadows in the room.

East Bay Hills House

EL CERRITO, CALIFORNIA
OHASHI DESIGN STUDIO

This home is a model of simplicity and restraint that uses all the right materials. Warm metallic tile frames the fireplace for the perfect modern detail. Ohashi Design Studio was given the challenge to update a 1960s ranch home with views of the San Francisco Bay. The use of tile to reflect the light from opposing windows creates an ever-changing focal point throughout the day. Its horizontal layout beautifully connects to the surrounding horizontal elements in the home, such as the siding shown on the entry above. The choice to recess the fireplace area adds dimensionality to the wall and provides an appropriate end-point for the tiles.

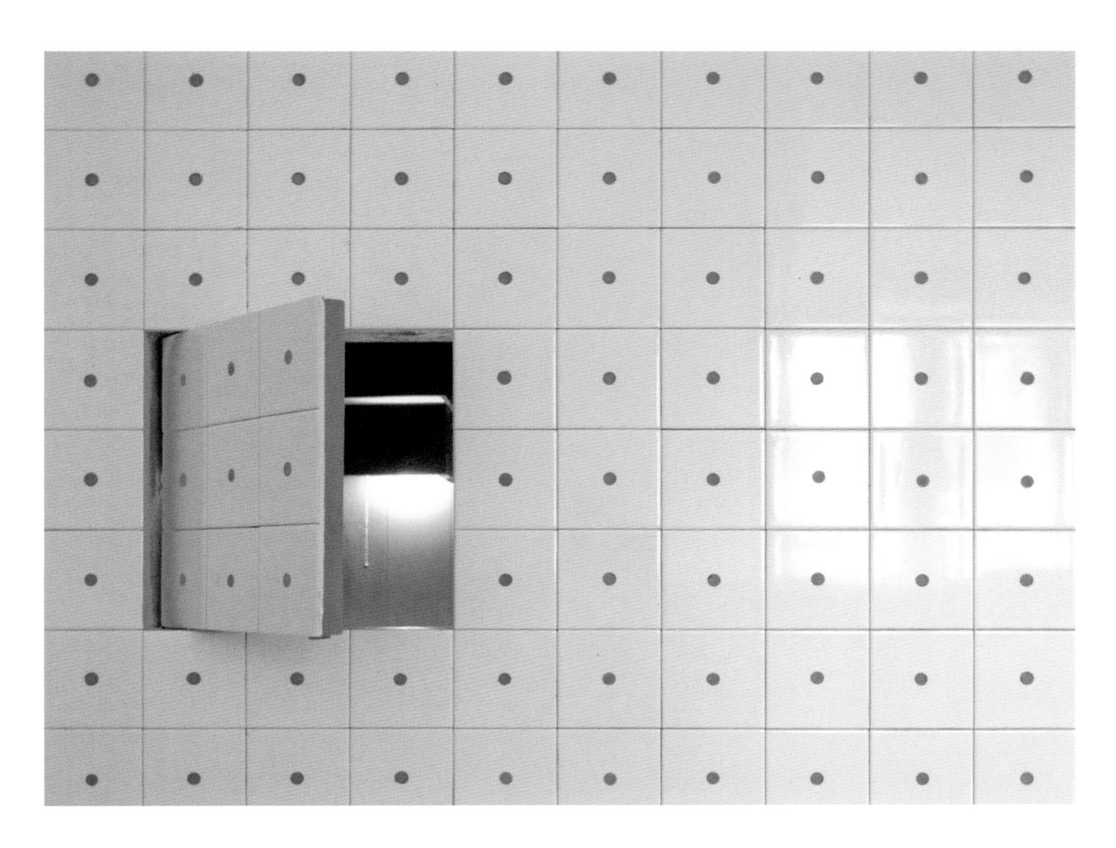

Hole in the Wall House

FÖHR, GERMANY

FRANCESCO DI GREGORIO AND KARIN MATZ

Here, the architects created two sleeping spaces in the hayloft of an old brick farmhouse in a maritime region of Germany where showing wealth meant tiling walls in unexpected rooms, in this case, the dining room. On their own, these simple square tiles, dotted with blue, are unusual but not necessarily remarkable. It's how they're used, all 3,200 of them, to create architectural volume, that makes them an inspired choice. The use of tile gives the walls solidity and presence that separates them from the drywall and paint that surround them. Their white glaze is a subtle departure from the rest of the painted surfaces, chosen to bring reflected light into a formerly dark attic. The blue dot sparks interest, begging for further investigation, which is rewarded after discovering the blue dot is a hole filled with blue cement that mirrors the blue ropes used to form a balustrade around the staircase. Its success is greater than if the architect had tried doing too much with a strong color or pattern—instead, simple materials with unusual detail are used to great effect.

Matbaren Bistro

STOCKHOLM, SWEDEN
STUDIOILSE

This bistro is the more casual of a pair of restaurants run by chef Mathias Dahlgren in Stockholm's Grand Hotel. The design starts with the food philosophy, which proudly boasts local heritage as well as new beginnings. This space, a food bar, was designed so one could visit comfortably on a regular basis for a quick drink or a light meal. The quality of materials, from wooden tables to the tiled floors, solidly expresses its functionality and robust à la carte menu. The earthy tiles maintain the natural feel of the wood paneling and tables while providing the one and only decorative pattern in the space. It's wonderfully refreshing to have such a strong visual interest on the floor, while the vertical surfaces and furnishings are more subdued and monochromatic. This allows the floor and all the surrounding elements to stand boldly on their own.

Vintage House

PORTLAND, OREGON
JESSICA HELGERSON INTERIOR DESIGN

This kitchen and dining room provide a wonderful example of how white tile is used to create a warm, cozy space. Lining the walls from floor-to-ceiling, the tiled areas include corners for verdant plants and cushioned sofas, resulting in a mix of textures, hard and soft. Tile gives the room solidity, framing the windows and the view outside, while the tiles' white glaze reflects exterior light, both soft and bright, which creates an open, light feel in a formerly dark space. Tiles placed around doorways give the openings strength and structure and contribute to a kitchen and dining area that double as a sanctuary. The alcove framing the range is particularly stunning, with custom trim finishing the opening, a deliberate and decorative choice that completes the space.

Lord House

SAN FRANCISCO, CALIFORNIA
JENI GAMBLE

Designer Jeni Gamble was challenged with redoing a poorly executed 80s-era remodel of a San Francisco Edwardian flat. She created a modern interior while paying homage to the original structure by using materials familiar to the era but with an updated, modern character. Ceramic tile plays a key role, as it represents a material that was originally common in homes of this era. In particular, Gamble chose hexagonal tiles for the bathroom, but in a larger format than one would expect to find in original Edwardian construction. She paired them with a traditional clawfoot tub, though in this case the tub is painted black. The good design comes from the consistency of the tile and color palette as one travels from room to room, noting a clean, modern look in familiar materials. Added details such as the subtle patterns of mixed gray and blue glazes on the floor, and only the floor, keep the look unique and original.

ZOMBIES
NEXT 5 MI

Neisha Crosland Studio

LONDON, ENGLAND
GREENWAY ARCHITECTS

This outdoor patio is brightened with hand-painted, glazed terracotta tiles designed by the studio's occupant, textile designer Neisha Crosland, who also designs wallpapers, rugs, vinyl flooring, scarves, and home office stationery. Stylized florals and geometric patterns inspired by nature; in bright but muted color combinations are her signature. The tiles fit in with the color palette of the otherwise spare outdoor space and its hedge of solid green, while providing the one focal point of decoration. The vibration of the patterns create a softness to the wall and bench below, in a great example of how a hard and solid material like tile doesn't always feel that way visually. The random but balanced placement of the different patterned tiles adds to this feeling. A border frames the top of the expanse of tiled wall, and no cuts were required, in an intentional act of planning to make the wall flow better as a singular piece, which is especially important for graphic tiles. More graphic tiles by the designer appear in the rest of the studio, which also serves as a showroom. A single pattern in a muted tone complements the mix of yellow brick in the small balcony, and another bolder graphic floor-to-ceiling pattern on an interior wall. Both provide a muted decoration, with modern takes on the traditional, and in contrast to otherwise spare spaces.

Chiselhurst House

LOS ANGELES, CALIFORNIA
BESTOR ARCHITECTURE

Each room in this Los Angeles home updated by Barbara Bestor seems to tell a different story, but from the same series. The oversize midnight blue tiled fireplace coupled with consistent green walls creates a unique and dramatic room out of a fairly undramatic rectangular space. The key is the contrast in the sheen of the glossy tile and the matte walls and moldings. The consistency in colors and large blocks of color also create a strong statement and avoid fussiness. The fireplace itself is a wonderful example of the richness one might not expect from simple, almost black, tile. The bright yellow bathroom tells a simpler story. The color and lively pattern in this room create a delightful place to start a day. The detail stopping the geometric tile pattern before the edges of the room creates balance, and the simple white tile used wall to wall and ceiling to ceiling create a solid and functional backdrop for the room. Like the living room, the bathroom keeps the count of materials and color to a minimum to striking effect.

Buena Vista House

SAN FRANCISCO, CALIFORNIA
GEORGE BRADLEY | ARCHITECTURE + DESIGN

For his own home, architect George Bradley and his partner Eddie Baba chose materials produced close to home first and let the design unfold from there, from the reclaimed redwood from nearby Hangar One at Moffett Airfield, to the Heath tile made less than two miles from their San Francisco home. The unique wood choice, which they used for cladding, ties the exterior to the interior, and the tile ties together the functional living areas, using variations of the same ceramic tile to create a holistic palette and a connected aesthetic. The centerpiece of the living area is a fireplace designed around three-dimensional oval tiles. The tiles are set vertically to accentuate the height of the space, while the texture of the tile minimizes the perceived mass of the fireplace, and a matte white glaze reflects the city lights from outside. The kitchen wall, clad in small diamonds, creates an integral yet subtle pattern, and the bathroom uses a fog blue glaze and full tile walls in a classic grid pattern.

Wanzenberg House

NEW YORK, NEW YORK
ALAN WANZENBERG

Architect Alan Wanzenberg has used Heath tile in his projects for years, including in his own homes. Given that he lives and works in Manhattan, many of the spaces are tight, including his own former apartment, and the use of tile is a more discrete element in the space. In the fireplace surround, the mix of red glazes he chose gives it a subtle vibrancy that sets it off from the neutral hues of natural woods, fabrics, and metals in the rest of the room. Beautiful materials and fine detailing are the common thread in every element and object in the space. The tile is no exception, with its subtly complex combinations of sizes and colors. The metal edging lends the whole fireplace the inlaid feel of a finely crafted jewelry box. It's a small expanse of tile that speaks volumes with the obviously careful and studious attention it's been given in its design. In his new house upstate, Wanzenberg had more space to use Heath tile but he still chose to use it like a jeweler in the detailed attention he gives each layout. When Alan found out that we weren't going to make the volcano red glaze anymore, he bought a few boxes in our classic oval tile and stashed them away for a spell. The result of this hibernation and rumination is the stunning wall over the kitchen sink (shown on the following page). The tiles are artfully placed in a setting (in this case, bands of painted wood and matching grout) to create a visually fascinating artwork. The blocking of the groups of tile with changes in orientation is a great example of what you can plan out when creating carefully intentional layouts. This wall contains many detailed blocks of interest. In contrast, the yellow laundry room tile is used in a more traditional way as a bright backdrop to a traditional farmhouse sink, including how it terminates at the washed out and simple wood planking that extends to the ceiling. The choice of color prevents it from feeling austere, giving it a touch of modernity, while still creating a clean space that's appropriate for a laundry room.

ONE GOOD DISH
super natural every day
the complete vegetarian handbook
THE ART OF SIMPLE FOOD
real food
BOTTOMFEEDER
AN EVERLASTING MEAL

HOW WE THINK ABOUT TILE

Why talk about *making* tile in a book on interiors? Because understanding tile, the various types, and the process of making them, can lead you to tile's multitude of aesthetic qualities and inform your choice of compatible materials and design elements.

Ultimately, it's about knowing your material. Factories and the fabrication are invaluable teachers for this. We have the luxury of the Heath factory, of course, but we love visiting others, seeing how their tile is made and how they create products with a unique aesthetic. From small to large, every tile producer does it a little differently, resulting in a product that's a little different, as well.

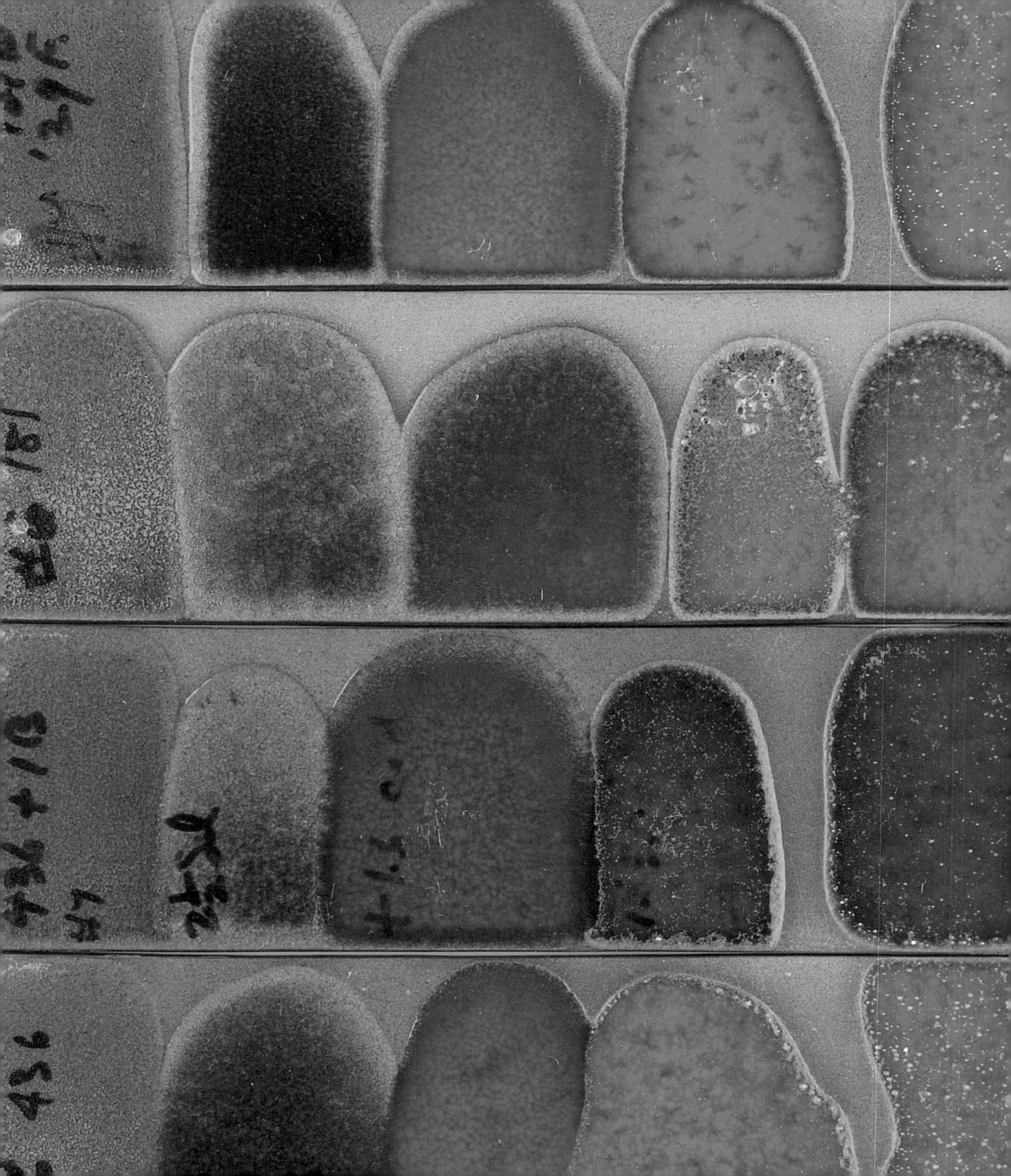

TYPES OF TILE

Good materials yield good products. And by good, we mean long-lasting, a little quirky, and slightly unpredictable—exactly how we like it at Heath. In order to better understand the final installation, how it lives in a space, and its design characteristics, let's start with its raw form.

At Heath, tile begins in our clay room: a dusty noisy space with mixing vats caked in 40 years of patina. A dry powder, which started as clay unearthed from a hole in the ground is combined with water and other minerals to start the clay-making process. The process is similar elsewhere, and there are generally three types of ceramic tile based upon the clay used:

1. Earthenware is a coarse reddish clay that's fired at low temperatures and produces the most colorful glazes. Earthenware tends to be the most handmade in feel, with thicker tiles and more irregular edges. This type of clay is porous, making it great for garden pots but not for showers. Associated with more traditional styles (think terracotta), it's probably the most common and has been made the longest—this was the clay used in early roof tiles and Islamic and Northern European tile making.

2. Porcelain, at the opposite end of the spectrum, is a very fine white clay that's fired at high temperatures. For the most part, other than cobalt blues, the firing burns out the glaze color. Porcelain tile is the finest, thinnest, strongest, and is also vitreous (impervious to water). It's good for all sorts of applications. Porcelain first originated in China, where the type of clay for porcelain was readily available.

3. Stoneware, what we make at Heath, is in the middle, providing solid strength along with a pretty great range of glaze colors. While it's fired hotter than earthenware, it's not so hot as to burn out the color. Stoneware can range from a more handmade feel to something more precise, depending on the forming process. It can be vitreous so also good for a broad range of uses. Stoneware makes good dinnerware and was the prevalent clay for this use in the early American pottery industry.

Not all tile is ceramic. Several of the preceding installations use concrete tiles, often with graphic elements, produced in a traditional, labor-intensive process. To create this colorful graphic surface, a thick layer of finer cement, sand, and pigments are arranged in a pattern mold and pressed into the surface of the base concrete, fusing the elements until they become one, so the graphics don't wear off over time. Heath's clay and glaze do something similar—they coalesce into one finished object, though through a very different process.

FORMING CERAMIC TILE

Tiles are either formed plastic or dry pressed.

Plastic refers to a wet lump of clay that's easily molded into a form without cracking or breaking—it's clay as you know it if you've ever taken a pottery class. The clay is then pushed into a flat shape, using one of several methods, and cut to the desired size. At Heath we extrude the clay through a narrow opening in a process that's similar to forming pasta but on a much larger scale (and not as tasty). Because of the force it takes to create a thin flat ribbon of clay, the tile tends to have a memory that results in surfaces that aren't quite flat—corners turn up a bit, and edges can turn out. Different clays and processes create more or less of these irregularities in the final tile, which becomes part of its individual aesthetic.

Dry pressing starts out with slightly damp powdered clay squeezed at high pressure to form a solid tile into a mold. Without the stresses of clay memory, where one shape has been forced into another, these tiles come out very flat, very precise and regular, and very consistent tile after tile. A box of dry-pressed tiles yields tiles that are generally identical in size and shape and are machine-made in appearance—think the kind of mass-produced tile you'd find for sale at large home improvement stores.

GLAZE AND SURFACE FINISH

Glazes are the glassy surface on a tile that give it its final color and finish. Though often, depending on its opacity, the color of the clay itself will show through the glaze. These surfaces come in a range of matte, shiny, or satin finishes, and are made up of a number of minerals and metals that define the color, opacity, and finish during their change of state in the firing process. Glaze chemistry and firing are not precise—ceramic engineers use a mix of chemistry knowledge, analysis, experience, and multiple trials to get it as close to a science as possible. So the next time you ask if that matte yellow can be done in a glossy glaze, the answer is likely no, as changing the finish will affect the color.

The level of a material's refinement affects the consistency and variability of a glaze, with a precisely engineered stain giving a more consistent result than the more capricious nature of a natural pigment. You're probably not shocked to hear, at Heath, we prefer the natural stuff, less refined raw materials—it keeps us on our toes, surprising and inspiring us with their sometimes unpredictable outcomes.

The method of glaze application is yet another factor that affects the aesthetics of a piece. Walk into Heath's glazing room (we give tours!) and you'll find individually manned glazing booths with spray guns. An automated "waterfall" glaze line (think what's for sale at big-box home improvement stores) will give a more consistent look than a hand-sprayed or hand-poured glaze. As with many other things, the outcome is quite different between high-volume mass-production and smaller artisanal processes.

As the final part of the process, tiles can be screened or printed to add graphics. A tile can be masked so that a glaze does not touch surfaces that define the pattern. Pad printing of a graphic transferred to the surface of the tile can also be done, to crisp or loose effect, with glazes or inks that don't burn out in the firing.

FIRING

The final process in making ceramic tile is firing it in the kiln. The tiles, clay body and applied glaze, are heated to a high temperature over a long enough period of time so that the materials melt, mature, and heal themselves into a glassy structure. Heath fires its kilns up to 2,080 degrees Fahrenheit over an eight hour firing, using pyrometric cones (top left) to determine when their contents are done firing. Cones measure the amount of heat absorbed. As the cone nears its maturing range, it softens and the tip bends, drawn down by gravity. Each higher cone number requires more heat to bend—Heath stoneware is fired to cone 01, porcelain to cone 10. The firing process is when the glaze gets its color and surface finish, and it's also when anything can happen. As much as you've tried to control all the factors, firing is where it all comes together. With that, one can intentionally leave room for surprises, depending on the aesthetic one is looking for. Our old, manually controlled kilns at Heath gave us a range of results even though we aim for consistency (there's a pottery tradition to blame such things on the weather, like the dampness or a particularly windy day). Our new computer-controlled kilns give us results not much different, which is why we still refer to our tile as perfectly imperfect.

Depending on how a tile is made and the materials used, one can get a very consistent product or one with lots of variability. It's opening a box in which each tile is identical, or each tile is different. The way we approach design is not from an individual tile's look and feel, but what all those tiles look like together on that wall or floor once the installation is complete.

INSTALLING

As good as any tile design may be, it's not done and it's not successful until it's installed with proper skill and intent. A curious thing we've learned over the years is that when an installation doesn't go well, the tile is blamed. Any material can be installed well or poorly, so it's about knowing how to make decisions and the questions to ask, long before you start installing.

You may love an aesthetic, but it may not work in your space. It's a skill to choose materials that fit, not fight, the overall design and construction of your space.

Any cuts to the tile ought to be considered in the design and, in particular, in the layout prior to the actual install. For a pattern to reveal itself, the scale of the installation space and the size of the room must work together. Grout color and width (both aesthetic and functional, depending on the material) play into pattern decisions at this time as well. Consideration of trims also fall into the aesthetic and functional category, as they can, and sometimes should be avoided, depending on the design.

Lastly, a tile installation starts way before you have the tile on hand. Tiling goes much more smoothly when your surfaces are flat and the dimensions are square, as both these will make a difference in the results. That said, work with what you have, respecting the space. It's also helpful to start working with a great tile installer early in the process to help make important decisions up front. We've learned through experiences good and bad that a really great installer is as much of a craftsman as he is an engineer, and this makes all the difference.

DESIGNING WITH TILE

Now that we've taken care of knowing our materials and knowing the different characteristics of tile, we can talk about using those materials, as well as designing them into an environment.

We begin by noticing the space—its textures, materials and presence, energy, lines, and feel. What is it asking for? Is it large or small? Public or private? Does it want to be bright or muted? Is its architecture modern or traditional? And on and on. We like things that feel intentional. In our case, because our home is a simple 1890s Victorian, there are tile designs we love that we would never use because they wouldn't be in keeping with the character or the intent of the design of the home. We've spent a lot of time resisting the urge to do any clean simple grids or a lot of strong colors (though we pushed it a bit with our yellow kitchen), and because we appreciate all sorts of architectural styles, we live vicariously through installations in other homes that are architecturally different from ours. This is one reason why the installations in this book are not only Heath tile, but span a broad range of tiles. Heath can be versatile, fitting into aesthetic styles from earthy to modern, and there are other great tiles out there that fit into other aesthetic niches.

Perhaps because we work directly with materials and the making process, we pay close attention to what goes into a product and allow it to inform its end use, including which materials it's mixed with. This has become our starting point for interiors and architecture. The quality of a material, from its raw form to how it's manufactured, is hard to hide. Tile will look like what it is—a thin, light tile will look that way installed, and heavier thicker tile will have a more solid appearance. A fired-on decal to create a pattern will not have the depth and richness of integral color. The quality of the starting tile will affect the quality of the final installation and the design of the space overall. The same goes for the other materials that will be part of the design of the space, even down to the paint. Ultimately, it's about whether materials have a soul and personality, or fall flat in their one-dimensionality. A consistency in the quality of all the materials makes a design more cohesive.

The size of a room and the space for tile are also very informative from the beginning. Pattern created with solid tile and grout lines, or with graphic tiles, typically need more space to have a strong effect as well as space to frame them. The size of the tile selected is also informed by the amount of space, as a large-format tile like 6 x 12" will have a very different feel in a small space than it would in an expansive room. One of the reasons we wanted to write this book is because we've seen more and more tile installations go beyond the usual functional borders to really define a space. These are the installations that cover a full wall and sometimes even wrap around the ceiling, tile around window frames to keep the material consistent, or tiles that extend the full wall behind open-backed kitchen cabinets and shelving. These installations make a real statement about tiles by highlighting their exceptional aesthetic qualitites beyond simply being a functional material.

01 COLOR

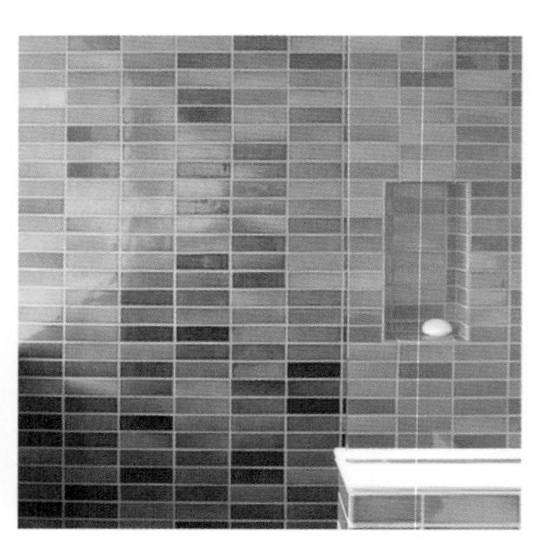

Color, viscerally and sensorially, defines a room's tone. Nothing else about tile makes as strong a statement, and that's precisely what makes choosing its color so exciting.

Making a bold choice takes confidence, no matter the installation size. Designer Merideth Boswell used Heath tile for nearly every surface of Clear Creek Spa, covering the floors, baths, walls, and ceilings in 2 x 12" green tiles in varying shades and finishes (left page, lower right). That's bold. And it worked!

Enter a room with a really good use of colored tile and you'll inevitably take notice. There is light and weight at play—texture, reflection, and the perceived permanence of tile—and that's exactly the reason tile is a more evocative material than paint.

Start by knowing the mood you wish to create and choose your materials accordingly. Take note: we find choosing a strong tile color makes pairing accompanying materials, finishes, and tones that much more important—they must work together, down to the furniture and appliances. That said, it's okay to play and even color outside the lines.

Truth be told, we haven't always agreed on color. We both lean toward the conservative, so imagine the discussion we had before tiling our kitchen in a very bright yellow. We went with it and we installed it ourselves. It was the boldest we'd ever been with color, and we love it to this day.

So give color some thought. A well-considered palette pulls together the entire home and keeps a sense of delight throughout.

02 WHITES

Admiring white tile—its strains and varieties—is a bit like admiring vanilla ice cream; deceptively simple, yet varied in its depth, richness, and texture.

Shortly after we bought Heath, designer Ann Sacks invited us to her home for brunch, excited to show off the classic white Heath tile in her beloved kitchen (left page, lower right). We enjoyed a wonderful meal, a beautiful home, new people, and an analogy that has stuck with us ever since: "White tile is not all the same," she told us. "A white rug made of fine wool is not the same as one made of acrylic fibers. No one would think those two are the same, and tile is no different."

In many ways, the study of white tiles tells one what's wonderful about ceramic tile as a material and design element. It acts as a canvas for the material's qualities. Artisanal tiles, like those from Heath, have inherent variation, where no two are the same. This variation creates an installation that feels warm, soft, and homey because of reflected light, surface finish, and texture. White on a flat, precise tile often feels stark and antiseptic, likely not something for the home (more like airport bathrooms and commercial kitchens).

Shades of white—warm to cool, and those with slight pigment—coupled with a range of surface finishes are what make white tile so rich in possibility. They range from traditional to minimal and modern to warm and inviting. The mix of white tiles in our own master bath (shown far left) does that, creating a space that feels light, clean and cozy. Plus it acts as an anchor for a whole floor where the prevalent color is white.

03 PATTERN

In writing this book, we had quite a discussion about what makes a pattern. Historically, patterned tile was an architectural detail used decoratively. Think highly intricate hand-painted tile in the Middle East and Spain, or inlaid tile with two types of clay and figurative reliefs in medieval times.

Interestingly, Islamic tile work became so ornate because religious beliefs forbade the use of imagery. Early styles were often graphical, and modern expressions of solid-colored patterns more understated. We focus on the modern form in this book, as it best fits our interpretation of what makes a pattern.

While our favorites subtly emerge from solid colored tiles, simply stacked, vertical or horizontal (sometimes both with a simple turn, see the Buena Vista House shower, page 203) we often see combinations outside this convention that surprise us, causing us to pause in admiration.

Almost every basic tile pattern has been done before, but it's the tile's color, size, positioning, and grout lines that create endless opportunity. Using a large tile to create a basic pattern (for example, running or Flemish bond) feels new and stunning, as these patterns are typically done using small tiles (for example, 2x4"). When designing a truly appealing installation, pattern repetition ought to be a major consideration.

And take grout seriously—it's not merely a finishing element. Its width and color are major factors in laying patterns. Grout has the power to tie together (or not) an installation and even star as the primary design element—if one wishes—in part because of spacing and in part because of color. The same tile laid with a wide grout line versus almost no grout will look completely different. And a high-contrast color of grout (to the tile) versus one that blends in are equally different in the outcome and impact of the installation.

04 TEXTURE

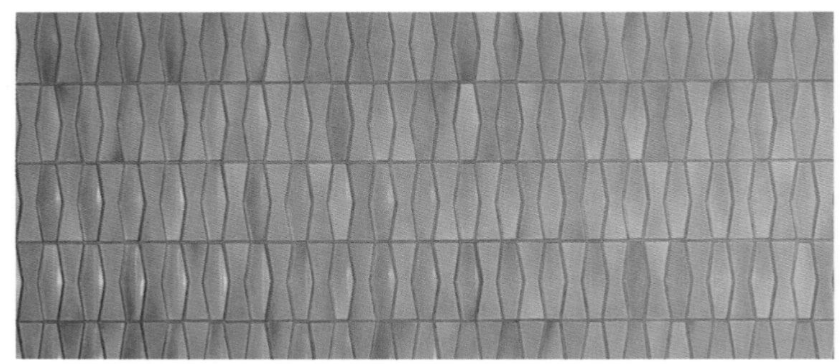

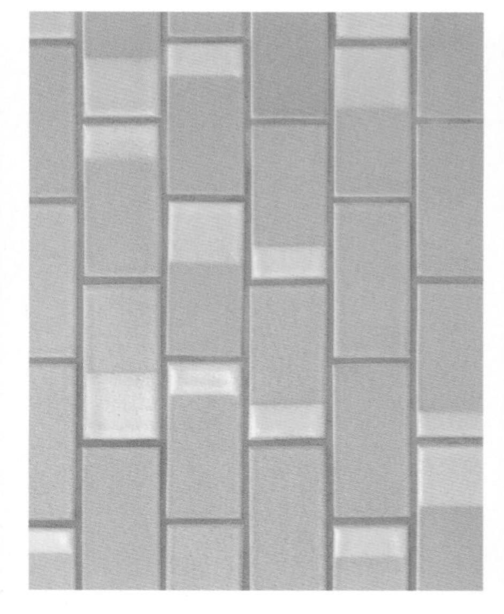

Texture aptly follows the discussion of pattern, since we think about texture as the surface of the tile, as well as its glaze.

Three-dimensional tiles, used decoratively and functionally, create stunning physical texture. These tiles, like Heath's ovals, have concave and convex surfaces and soft and hard edges. They're more than just slightly raised, as adding more complex three-dimensionality can change an installation quite a bit, making it far more textured than an individual tile suggests. So, a three-dimensional tile can be installed in a pure way that lets the shape stand out, or the finished installation can be a result of new three-dimensionality that results from a grouping.

Mixing finishes—matte, glossy, satin—is another way to add texture, using reflectivity for subtlety and elegance or captivating strength. Our bathroom, shown above, exemplifies this, using whites in varying finishes for a warm, quiet installation. This is equally stunning in black, as in the fireplace wall that Cathy installed in our home (see page 21). With whites and blacks, the effect comes more from the finishes, and with colors, tone and finish create an installation that changes with the light.

High variation in the glaze itself creates a subtle and beautiful texture. We love and embrace variation at Heath, even using raw materials that encourage it. Early on, variation in glazes taught us about an installation coming together as a whole, rather than it being about an individual tile—you don't learn that from buying white tile off the shelf at the building supply store.

We encourage distributing variation throughout an installation, but we've even seen designers use variation within a glaze to create a fade and even a blocking pattern. Success depends on how hands-on you, the designer, want to be. And we mean really hands on, as in laying out the entire installation on the floor before it gets put into place on a wall. Great results require getting on your hands and knees, and great designers wear knee pads.

05 GRAPHIC TILE

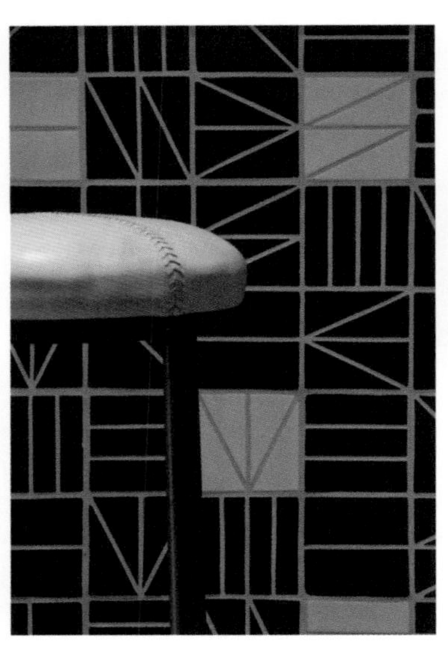

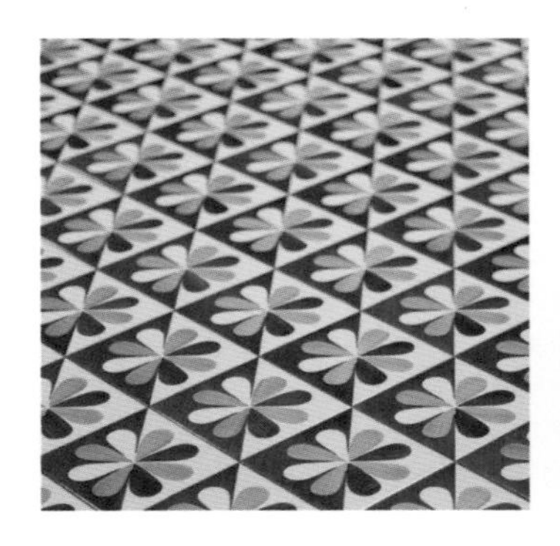

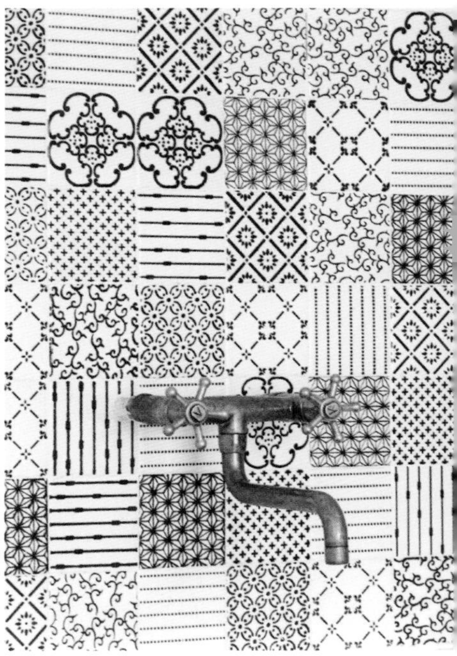

Innately more decorative and less modern, graphic tile was born to tell stories.

In traditional Spanish-style homes, graphic tiles enhance the history and appropriateness of the architecture. That's not to say the home can't feel modern, but an accompaniment of traditional tile grounds the design and tells the story of its beginning.

Some of the installations we chose for this book use graphic tile playfully. Playful is good, particularly with a material as substantial as tile, as it shows confidence and commitment.

Today, traditional Mexican and Moroccan graphic tile is typically found in spaces with an aesthetic that is recognizable as regionally specific. However, graphic tile is also used well in contemporary spaces to provide simple silhouettes of color that add texture and depth to a space, offering a permanence unlike most other materials.

Modern graphic tile can also be a backdrop for contemporary fireplaces or wood stoves—a reference to traditional installations found in Northern Europe in the last century that were clad in tile for functional reasons, often with traditional graphic elements or pictures.

Graphic tile can also feel soft and take the visual place of a rug or textile. In fact, we've seen some great floor tile installed in the style of a rug (see Parco dei Principi Hotel, pages 104-109), and, in one case, even a doormat, that we enjoy for its tongue-in-cheek spirit. It's also a great reminder that tile can be used as purely decorative, as well—the São Paulo House (pages 160-163) made us smile for this reason. Is tile the new wallpaper?

06 ARCHITECTURE

There used to be fantastically complex ceramic elements on the exteriors of buildings built in America in the first half of the 1900s.

The Eastern Columbia Building in downtown Los Angeles is one such building. Many purely decorative elements were custom-made by companies like, Gladding, McBean of California, which employed talented artists and craftspeople for custom commissions. Today, the work is mainly restoration and renovation of historic buildings, not the enhancing of new ones.

It wasn't uncommon for entire facades to be clad in ceramic material. One of the first major installations of Heath tile was the exterior of the Norton Simon Museum in Pasadena, California. We visited this building shortly after buying Heath. Its exterior is clad in volcanic colored tile with a textured glaze that gives a monolithic presence, with a soft skin that absorbs and reflects the Southern California light.

The tiles are still there today (and look fantastic), as tile lasts in a way that other materials don't—the sun won't fade it and the rain won't damage it. Used architecturally to define a form, it provides both softness and color along with permanence.

Sadly, budgets rarely allow entire buildings to be clad in tile any longer. However, there are some great examples of tile being used architecturally in this book, on a smaller scale.

The tiled interior of the Hole in the Wall House (pages 182–185) gives us inspiration. It's playful and we can't imagine how another material could do what the tile is doing while also giving the architectural elements weight. In this case, as in others, the tile gives one the sense that this element of the architecture is meant to stay there for a long time. This reiterates a consistent theme in our lives: we enjoy the spirit of things with intention.

07 ART AND EXPRESSION

Clay has been around forever, as both a functional and artistic medium.

Because of its durability, clay is an archaeological artifact that tells us a great deal about ancient civilizations—how it was used functionally and artistically tells a cultural story. Tile is actually one of the more functional offshoots of clay that quickly became decorative when makers began using and experimenting with glazes.

This book is very much about tile as a functional material and the artistry in its final form, rather than in an individual piece (you just don't see intricate hand-painted patterns and pictures the way you used to). We get pretty excited when a ceramic tile is used as a building material and an overt expression of art.

This type of custom work is rare these days, but there are a few artists that still work in this medium—Stan Bitters being one and Yellowwoods Arts in South Africa another. We're fortunate to have gotten to know Stan through a gallery show at Heath San Francisco. Stan has been experimenting with ceramic sculpture and glaze since the 1960s. In the installations featuring his work in this book, it has been integrated into large-scale tiles that serve as cladding for vertical surfaces and turn walls into monumental sculptures (left page, middle).

Other installations in this category use painted tile. Glazes create decorative patterns or pictures much like the way decorative tile evolved originally but, of course, with modern themes.

Tile need not be custom to create artistic installations. One of our favorite methods is simply using tile unconventionally, outside the way it's meant to be used, to create fully encompassing environments—because this is what art does in this medium, and sometimes it crosses over into architecture. Giò Ponti's Parco dei Principi Hotel on Italy's Amalfi Coast (pages 104–109) is a wonderful large-scale example of this. There are smaller examples that can be just as delightful, such as tile in places one wouldn't expect: like ceilings and stair risers.

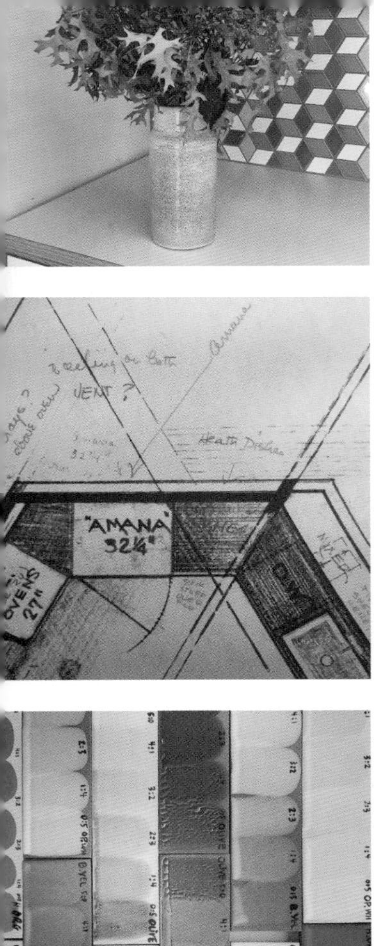
"AMANA"
32¼"
OVENS
27"

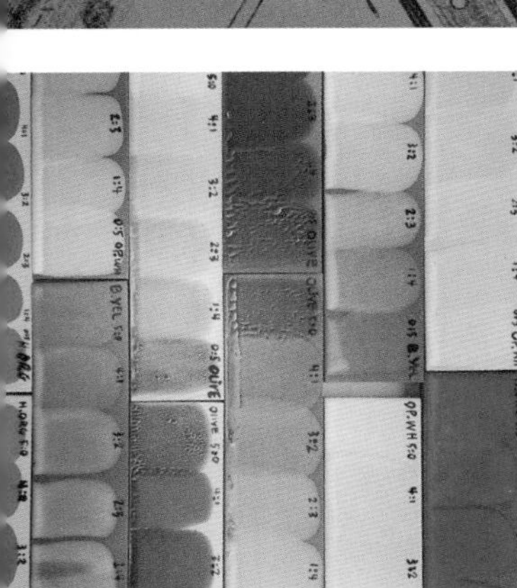

B.YELLOW
PERSIMMON

HEATH

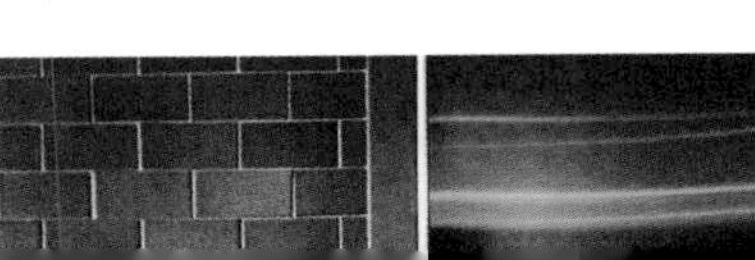

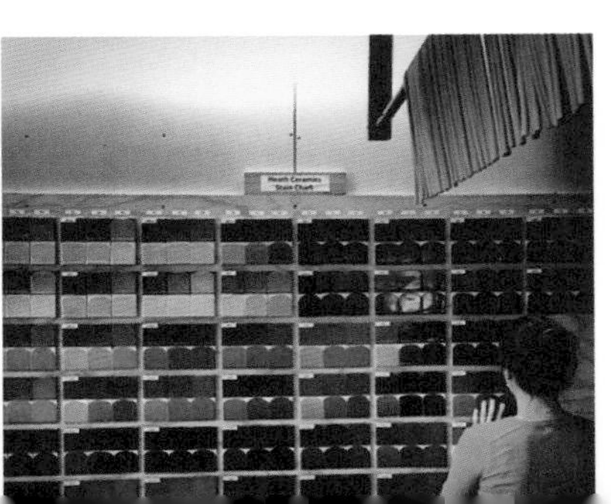

ENERO
2013
FEBRERO

HEATH

SAN FRANCISCO

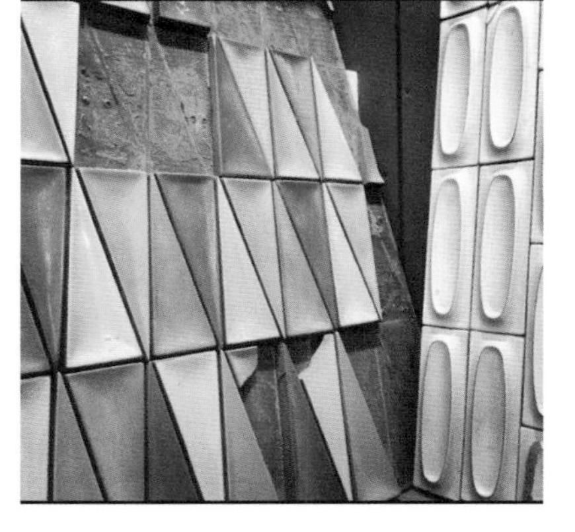

HEATH

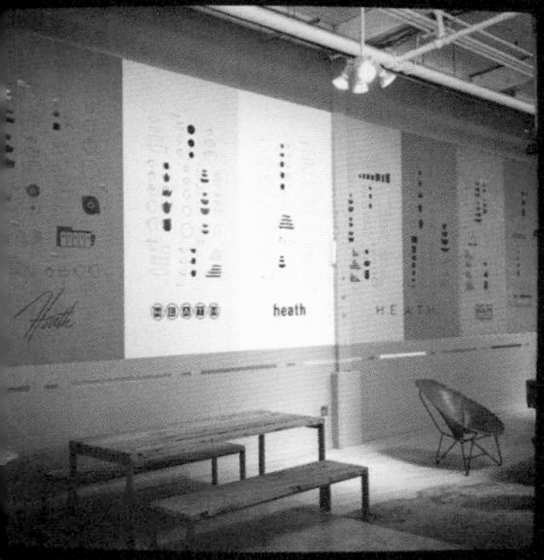
heath

HEATH

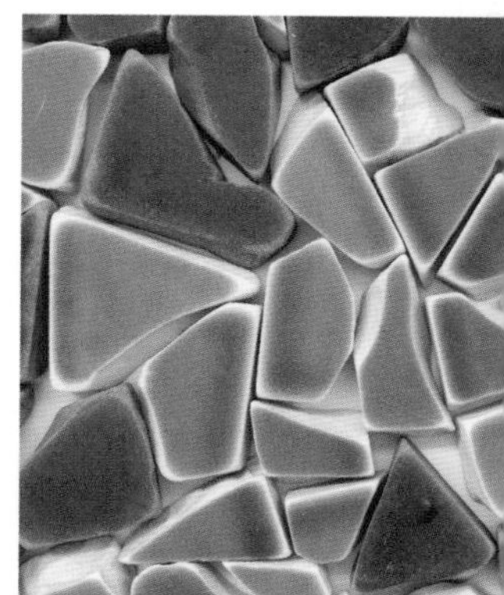

HEATH

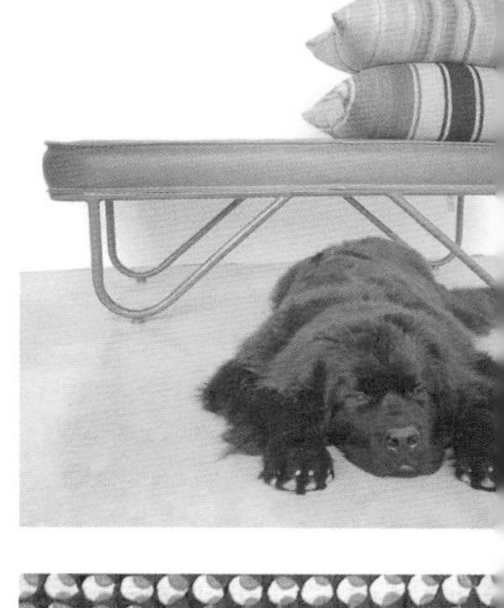

TILE TRAVEL

We're surrounded by great public tile installations. Like the architecture it accompanies, tile's aesthetic and how it's used tells a story of the country and place. When we travel, we make a point of visiting any significant installation. Outside of tile we know about, our eyes are always open to surprising and inspiring displays. The recommendations below are by no means complete. The selection ranges from rich, cultural, and historical to singular and artistic to notable and classic and finally, to innovative and modern. Should you want more, we've thrown in a couple of places where you can dig deeper to see how tile is made. As with any form of inspiration, it helps to leave the comfort of your home and see the world, in all its glory.

Downtown LA

LOS ANGELES, CALIFORNIA

Downtown L.A. recently underwent a cultural revival. With more reasons to travel on foot, there are more opportunities to look up, take notice, and be amazed by the ceramic tile that clads some of the deco-style buildings. The Eastern Columbia is one such reason. As we walked past it one day, we first noticed tile just above the street level. We looked upward at the tower and were blown away by the sea of turquoise blue glaze and the full extent of tile work. It's a bit of an exercise to find the right views, but this and the exterior of many other buildings in the neighborhood are worth it.

South Broadway St at West 9th St., Los Angeles, California

Brasília

BRASÍLIA, BRAZIL

Designed in the 1950s by architects Lúcio Costa and Oscar Niemeyer as Brazil's new capital, Brasília is known for its grand modernist architecture and utopian city plan. Works by incredible artists and sculptors adorn public spaces. From bus stops to the interiors of grand city buildings, Brasília provides more than a few opportunities to see tile. The tile of Athos Bulcão, a notable artist, can be found throughout the city: the Brasília International Airport, the Cathedral of Brasília, the National Congress, and many more delightful locations. Also worth noting, tile by the Athos Bulcão Foundation can be seen lining the swimming pool wall at the Fluid House (pages 54-57).

arcspace.com/travel/travel-guide-brasilia/

National Tile Museum

LISBON, PORTUGAL

There's a strong cultural component to this museum, showcasing the making of azulejos tile, its form of expression, and its rich history in the region. With Moorish influence, the azulejos, or tin-glazed ornamental tile, has become a Portuguese art form. The collection spans the history of tile from the fifteenth century to today. As the name, National Tile Museum, suggests, the exhibits are exclusively tile. So if you are a tile lover, be prepared for total immersion.

www.museudoazulejo.pt

Swedish tiled stoves

SWEDEN

Traditional Swedish tiled stoves, often painted in decorative patterns and featuring fancy ceramic medallions, were once the prevalent source of warmth during long Scandinavian winters. These stoves are reputed for efficiency, in part because they're lined inside and out with porcelain tile that retains the heat, slowly releasing it. We were taken by the first one we saw, in Danish floral designer Tage Andersen's old mansion on the grounds of his museum and gardens at Gunillaberg, Sweden (well worth the visit!). We later found the aficionado's destination for these functionally beautiful objects is just a few hundred kilometers away at the Swedish Porcelain Store Museum.

tage-andersen.com/gunillaberg_uk.html
www.kakelugnar.org

The Alhambra

GRANADA, SPAIN

We both, separately, visited the Alhambra in our early travels around Europe, well before we ever imagined we'd be designing and making tile. It's unlikely the reason we were there was because of tile, but its ornate detail, color, and pattern stuck with us long after we returned. The castle dates back to ninth century AD, while its Moorish palaces were built over the next several centuries. The style of architecture seen today, attributed to the Nasrid Dynasty, left no surface without decoration. Now a museum, visiting is a rich experience from the perspective of both historian and tile lover.

alhambradegranada.org

Dome of the Rock

JERUSALEM, ISRAEL

This shrine, one of the oldest examples of Islamic architecture, glistens in color on the Temple Mount, a historically and religiously significant area of Old Jerusalem. The tiled facade gives the building its brightly colored appearance, with rich blues, greens, and golds prevalent in the mix of ornately patterned tile. Approximately forty-five thousand tiles line the exterior of the building and extend to the columns and ceilings of the arcade, most being installed around the year 1500AD. The tile work is a wonderful example of Islamic art, and more elaborate tile is used to decorate the interior. However, though the exterior can be accessed by visitors at certain times, access to the interior is very limited.

domeoftherock.net

Moravian Pottery and Tile Works

DOYLESTOWN, PENNSYLVANIA

A National Historic Landmark, the museum still produces handmade reproduction tile in the same designs and using similar processes as when it was fully functional as a tile factory. The pottery, and its founder Henry Mercer, were significant to the United State's Arts and Crafts movement. Founded in 1898, its original buildings are intact and are as much of a reason to visit as anything else. The museum features historical exhibits, installations, and examples of tile made on-site, as well as production processes for making its decorative tile. If you have the time, you can take one of its classes to learn how to make one yourself.

buckscounty.org

Blatt Chaya Tile Company

BEIRUT, LEBANON

This family business, started in 1881 and relaunched after a 60-year hiatus in 1996 by the same family, makes traditional inlaid cement tile. The father, who brought Blatt Chaya out of retirement and has the focus of an old-school craftsman, runs the company together with his daughter and son (who studied design at Rhode Island School of Design). Making the traditional cement tile is a laborious, handmade process the owners proudly tout, with their stand off against, what they call, the disease known as speed. The workshop can be visited for tours.

blattchaya.com

Heath Ceramics Tile Factory

SAN FRANCISCO, CALIFORNIA

Our tile factory, built in 2012, like our dinnerware factory in Sausalito, is open for tours. We make stoneware tile, extruded and cut to size, dried, glazed, and fired in large "top hat" kilns. Our processes are a mix of modern machinery and hand craftsmanship. You can book a tour on our website.

heathceramics.com

Osaka Aquarium

OSAKA, JAPAN

If you're not with young children or you're not a lover of marine life, there may be little reason to visit this aquarium, save the incredible tile installation cladding its front. We've noticed a small worldwide trend: aquarium architecture and tile murals. The Osaka Aquarium is one of the largest in the world, and features an impressive five-story tiled facade with depictions of sea life, created by architect and exhibit designer Peter Chermayeff. It's playful, colorful, and very representative of 1980s tile design, but the geometric patterns and overwhelming scale of the tile on public architecture of this size make it worth standing under its deep blue tile sea.

kaiyukan.com

Lisbon Oceanarium

LISBON, PORTUGAL

The second tiled aquarium on our tour, an original, features a stunning six-story, 240-foot-long, mural called *Tiles of the Oceans*, billed as the largest mural in the world. Also by architect Peter Chermayeff, it's layered in complexity. Up close, you see microcosms of marine life. Far away, fifty-four thousand tiles change incrementally from white to dark blue and weave through the interior and exterior of the building. The pixilation creates a shimmery effect, cleverly depicting an underwater scene. It being Portugal, the tile continues beyond the mural. A new addition to the aquarium features a facade of five thousand custom-made white tiles representing fish scales.

oceanario.pt

Parco Dei Principi Hotel

SORRENTO, ITALY

This is not only an opportunity to see great tile, but to be immersed in it, if you choose to be a guest at the hotel (or even dine at the restaurant). It was designed in 1962 by Giò Ponti, one of Italy's greatest modernist architects. Ponti designed the interiors and architecture, as well as thirty graphic tiles that vary in application and pattern throughout the hotel. The concept of tile goes further, with ceramic pebbles set into the reception desk and other areas. Sculptor Fausto Melotti also created tiles for the dining room walls. The hotel remains a popular destination in Sorrento, Italy, for good reason.

royalgroup.it/parcodeiprincipi/en

Roberto Burle Marx

RIO DE JANEIRO, BRAZIL

The legendary work of landscape architect and artist Roberto Burle Marx, often credited for introducing modernist landscape architecture to Brazil, can be admired in many public spaces throughout the country. In Rio de Janeiro, the Instituto Moreira Salles, now a non profit cultural center open to the public, features an undulating wall that wraps around the pool. In various shades of blue, the tiles are hand-painted and depict washerwomen in an abstract cubist style. Also, the incredible Copacabana Promenade, a four-kilometer stretch of paved mosaic landscape in a geometric wave, was completed in 1970.

ims.com.br

Casa Batlló

BARCELONA, SPAIN

Built in 1906 as a private residence, this house by Antoni Gaudí recently went through a major conservation project and and is now open to the public. Covered in shifting colors of mosaic tile with skeletal forms, the home's nickname is "the house of bones." While tile is used throughout (the interior is equally amazing), it culminates on the roof with a section shaped like a dragon's back, fantastical spires that terminate the chimney stacks and turrets, all of which are tiled in brightly colored mosaics that are undoubtedly the work of Gaudí.

casabatllo.es/en/visit/

Adamson House

MALIBU, CALIFORNIA

This historic home, a Spanish Colonial Revival designed by architect Stiles Clement in 1929, is now open to the public. A visit provides a glimpse of both California and tile history. The house was built by Merritt Adamson and Rhoda Rindge, who's mother May Rindge started the famed Malibu Pottery (responsible for some of the most beautiful tile of the era). Extensively tiled both inside and out, the home is an amazing tribute to Malibu Pottery (which was located less than a mile away). The pottery burned down in 1932, amazingly producing a legacy after only six years. Today, it's the de facto site for fans of Malibu and this era in California tile making.

adamsonhouse.org

Duncan Ceramic Products

FRESNO, CALIFORNIA

The exterior of this building, the headquarters for a ceramic supplies company for the hobby market, features massive ceramic sculptures and murals by ceramic artist Stan Bitters. A Fresno native who still lives and works in a studio there, Bitters created these pieces in his signature bright glazes and Brutalist style in 1969. The environmental scale of these pieces is truly another way to experience ceramic art and makes for one of the more inspiring parking lots of any office building anywhere. Bitters even made the company sign.

stanbitters.com/architectural-installations

Norton Simon Museum

PASADENA, CALIFORNIA

Blown away by its beautiful and soulful use of tile, this was the first building we visited after becoming the owners of Heath. Opened to the public in 1969, the curving exterior walls are clad entirely in tiles designed by Edith Heath, all 115,000. The unusual size, 5 by 15" is complemented by a glaze of deep volcanic umber that vary greatly in color and sheen. The light reflecting on the dark hue creates a feeling that's at home in the building's park-like surroundings. Along with this unusual and inspiring architecture, the intimately scaled museum also houses a wonderful collection of European and Asian art stemming from the private collection of twentieth-century industrialist, Norton Simon.

nortonsimon.org

Sydney Opera House

SYDNEY, AUSTRALIA

The iconic building by architect Jørn Utzon, completed in the early 1970s, is notable for the unique shape of its roofs, affectionately called sails, which are covered entirely in ceramic tile. Made by Swedish ceramics company Hoganas Keramik, an impressive total of 1,056,006 tiles in glossy and matte white create a reflective and textured effect that changes with the strong sun of Sydney's harbor. To upkeep the building, now more than forty years old, the Opera House developed a program where one figuratively "sponsors a tile" to help ensure that the beauty of this amazing structure continues into future decades (and we hope it does!).

ownourhouse.com.au and sydneyoperahouse.com

Santa Caterina Market

BARCELONA, SPAIN

Originally built in 1848, this food market was recently updated in 2005. Architects Enric Miralles and Benedetta Tagliabue considered the surviving old structures as useful elements in building something new, revitalizing a neighborhood and moving it forward without tearing it down. The market is now a center of vibrant food culture. The tastes and smells inside the building will consume you, though you can't miss the massive undulating roof above, covered with 325,000 colorful ceramic tiles by artist Toni Cumella. While this delightful tiled work can be best viewed from the surrounding buildings, a satisfying peek can be had from the street at the front of the market.

mercatsantacaterina.com

PROJECT DETAILS

01. **Farmshop Restaurant**
Pages: 48-51
Place: Walls and counter fronts in a restaurant in Larkspur, California
Tile: page 48-51—Heath Ceramics stoneware 2x4" Field Tile in Olive #3 (heathceramics.com); page 50-51—Heath Ceramics stoneware 2x4" Field Tile in Mid Century White (heathceramics.com)
Date: 2013
Other Materials: Page 48—Tables and Bar-Top by Alma Allen (almaallen.com), Stools in leather and walnut custom by Commune (communedesign.com); page 49—on left, Arrowhead Lounge Chair by Plane Furniture (planefurniture.com), Commune Crate by Environment (environmentfurniture.com), Coffee Table by Espenet Furniture
Designer/Architect: Commune Design—Los Angeles, California and communedesign.com
Photographer: Mariko Reed—marikoreed.com

02. **Emery House**
Pages: 52-53
Place: Kitchen backsplash and wall in a private home in Brussels, Belgium
Date: 1996
Tile: Zelliges hand-cut tiles from Fez, Morocco by Emery & Cie (emeryetcie.com)
Other Materials: Page 36—Teapot found at a Brussels flea market; page 37—Star Chandelier found in the Souk in Marrakech, Morrocco
Designer/Architect: Agnes Emery
Photographer: Morten Holtum—holtum.dk; styled by Lykke Foged—lykke-foged.dk

03. **Fluid House**
Pages: 54-57
Place: Outdoor wall in a private home in São Paulo, Brazil
Date: 2013
Tile: Custom tile by Athos Bulcão Foundation (fundathos.org.br)
Other Materials: Wood decking is Cumaru (Brazilian Teak)
Designer/Architect: CR2 Arquitetura—São Paulo, Brazil and cr2arquitetura.com.br
Photographer: Rafaela Netto

04. **Nichols Canyon House**
Pages: 58-63
Place: Outdoor counter, showers, and outdoor sculptures in a private home in Los Angeles, California
Date: 2010
Tile: Page 58,60—Heath Ceramics stoneware 2x4"-Field Tile in Matte Brown (heathceramics.com); page 59—Heath Ceramics stoneware 2x4" Field Tile in New Gunmetal (heathceramics.com); page 63—Heath Ceramics stoneware 3x12"-Field Tile in Antique White, Mid Century White, Stone White, and Opaque White (heathceramics.com); page 61,62—Ceramic tile murals by ceramic artist Stan Bitters (stanbitters.com)
Other Materials: Page 63—Bathtub by Boffi in white Corian (boffi.com); page 58, 60—Bar stools by Harry Bertoia for Knoll (knoll.com); page 60-61—Chaise loungers and side tables sold by Espasso (espasso.com)
Designer/Architect: Commune Design—Los Angeles, California and communedesign.com
Photographer: Mariko Reed—marikoreed.com

05. **Murnane House**
Pages: 64-65
Place: Bathroom floor in a private home in Los Angeles, California
Date: 2012
Tile: Page 64, 65—Dandelion Encaustic Cement Tiles by Claesson Kovisto Rune (contemporarytiles.se); page 64—wall tile is Thassos Marble
Other Materials: Bathtub is AMAZE oval by Produits Neptune (produitsneptune.com); Fixtures are Avalon Collection from California Faucets (calfaucets.com)
Designer/Architect: Project M+—Los Angeles, California and projectmplus.com
Photographer: Mimi Giboin—mimigiboin.com

06. **Cristalli Bath**
Pages: 66-67
Place: Bathroom floor and shower walls in a private home in Copenhagen, Denmark
Date: 2011
Tile: Made a Mano Cristalli lava stone tiles (madeamano.com)
Designer/Architect: Made A Mano with Dorte Høegh—Copenhagen, DK and madeamano.com
Photographer: Heidi Lerkenfeldt—lerkenfeldt.dk

07. **Studioilse**
Pages: 68-71
Place: Communal kitchen in a design studio in London, England
Date: 2014
Tile: 4x4" Field Tile in Antique White by Koninklijke Tichelaar Makkum (tichelaar.com)
Other Materials: Page 70-71—wooden table and settle by Stuidoilse for De La Espada (delaespada.com), Bumling Pendant lamp by Anders Pehrson for Ateljé Lyktan (atelje-lyktan.se)

Designer/Architect: Studioilse—London, England and studioilse.com
Photographer: Mariko Reed—marikoreed.com

08. **Catalina House**
Pages: 72-77
Place: Exterior floors and walls in a private home in Los Angeles, California
Date: 2009
Tile: Page 72-73—Mosaic fountain backsplash, no longer available; page 72, 76, 77—floor tile is custom made cement tile by Commune; page 74—Patio floor tiles are Terracotta from Mission Tile West (missiontilewest.com), Terracotta wind chime screen vintage by Stan Bitters (stanbitters.com); page 75—Fireplace tile is vintage
Other Materials: Page 74—Croissant Sofa by Kenneth Cobonpue (kennethcobonpue.com), side table is vintage by David Cressey; page 76—Outdoor furniture is vintage 1940s French, coffee table with bronze legs and travertine top by 10Ten (ten10site.com); page 77—Stools are vintage Van Keppel-Green
Designer/Architect: Commune Design—Los Angeles, California and communedesign.com
Photographer: Mariko Reed—marikoreed.com

09. **Hotel Okura**
Pages: 78-79
Location: Interior walls in a hotel in Tokyo, Japan
Date: 1962
Tile: Record of the tile manufacturer not available
Designer/Architect: Yoshiro Taniguchi, Hideo Kosaka, Hajime Shimizu, Akira Iwama, Kisaburo Ito
Photographer: Aya Sunahara

10. **Stand-Alone Bath**
Pages: 80-81
Place: Bathroom and shower module in a private home (former stable) in Parma, Italy
Date: 2011
Tile: White 4×4" ceramic tiles by Cotto Veneto (cottoveneto.it)
Designer/Architect: Francesco Di Gregorio—Parma, Italy and francescodigregorio.it; Karin Matz—Stockholm, Sweden and karinmatz.se
Photographer: Francesco Di Gregorio

11. **Wedel House**
Pages: 82-83
Place: Kitchen backsplash in a private home in Sacramento, California
Date: 2013
Tile: Heath Ceramics stoneware 5" Hex in Campari Red (heathceramics.com)
Other Materials: Page 82—Range by Bertazzoni (bertazzoni.com), found branch shelf brackets by Nature
Designer/Architect: Popp Littrell Architecture + Interiors—Sacramento, California and plarch.com
Photographer: Mariko Reed—marikoreed.com

12. **Heath Ceramics Los Angeles**
Pages: 84-85
Place: Walls in showroom in Los Angeles, California
Date: 2008
Tile: Page 84—Heath Ceramics stoneware 3x12" in Opal Blue, New Sky Blue, Bay Blue, And Midnight #4 (heathceramics.com); page 85—Heath Ceramics stoneware 2x4" Field Tile Dual Glaze in Paprika (heathceramics.com)
Other Materials: Page 84—Pendant lamps by Adam Silverman (adamsilverman.net), Commune Crate by Environment (environmentfurniture.com), furniture is vintage; page 85—Vases by Adam Silverman (adamsilverman.net)
Designer/Architect: Commune Design—Los Angeles, California and communedesign.com; Catherine Bailey
Photographer: Mariko Reed—marikoreed.com; Corey Walter—coreywalter.com

13. **Heath Ceramics San Francisco**
Pages: 86-89
Place: Walls in a design studio, showroom, café, and breakroom in San Francisco, California
Date: 2012
Tile: Page 86—Heath Ceramics stoneware Mural Pattern in Arctic (heathceramics.com); page 87—Heath Ceramics stoneware Crease 3x9" in Frost on foreground counter, Heath Ceramics stoneware 3x12"-Field Tile in Opal Blue, New Sky Blue, Bay Blue, and Midnight #4 on background wall (heathceramics.com); page 88—Heath Ceramics stoneware 4x4" experimental screen printed tile; page 89—Heath Ceramics stoneware 4x8" and 4x4" Field Tile in Shark, New Slate #2, Shade Light, Blue Lake, and 4x4" Field Tile in Shade with custom pressed graphic by Geoff McFetridge (heathceramics.com)
Other Materials: Page 86—vases by Tung Chiang (heathceramics.com); page 87—stoneware serving bowls by Heath Ceramics (heathceramics.com), custom cabinetry designed by Commune Design; page 88—6 burner range and oven by American Range (americanrange.com)
Designer/Architect: Commune Design—Los Angeles, California and communedesign.com;

Catherine Bailey and the
Heath Ceramics Design Team
Photographer: Mariko Reed—
marikoreed.com

14. Beachwood Café
Pages: 90-91
Place: Floors in a café in Los Angeles, California
Date: 2012
Tile: Concrete tile 8x8" Khufu pattern by Granada Tile (granadatile.com)
Other Materials: Page 90,91—"Flowers" wallpaper by Geoff McFetridge and Pottok Prints (pottokprints.com); page 91—Vintage bent plywood chairs from Amsterdam Modern (amsterdammodern.com)
Designer/Architect: Bestor Architecture—Los Angeles, California and bestorarchitecture.com
Photographer: Mariko Reed—
marikoreed.com

15. Popp House
Pages: 92-93
Place: Shower in a private home in Sacramento, California
Date: 2010
Tile: Heath Ceramics stoneware 5" Hex in Tropics Blue (heathceramics.com)
Designer/Architect: Popp Littrell Architecture + Interiors—Sacramento, California and plarch.com
Photographer: Mariko Reed—
marikoreed.com

16. Upstairs Office and Apartment Heath San Francisco
Pages: 94-97
Place: Kitchen backsplash in an office at Heath in San Francisco, California
Date: 2014
Tile: Page 94,95 Heath Ceramics stoneware 4x4" Field Tile in Layered Glaze Chalk (heathceramics.com); page 96-97—Heath Ceramics stoneware Mural Pattern in Rhythm (heathceramics.com)
Other Materials: Page 95—4 burner range and oven by American Range (americanrange.com), Flour Sack tea towel by Heath Ceramics and House Industries (heathceramics.com); page 96-97—Wall unit and chair are vintage, Reese Sectional by Room and Board (roomandboard.com)
Designer/Architect: Catherine Bailey and the Heath Design Team
Photographer: Mariko Reed—
marikoreed.com

17. Butterfly House
Pages: 98-101
Place: Shower walls and kitchen backsplashes in a private home in San Francisco, California
Date: 2013
Tile: Page 98—Heath Ceramics stoneware 5" Hex in Stone Gray (heathceramics.com); page 99—Heath Ceramics stoneware Wide Hex in New Sky Blue (heathceramics.com); page 100-101—Heath Ceramics stoneware 3x9" Field Tile in Chalk White (heathceramics.com)
Other Materials: Page 98,99—Shower walls and floors are custom cast concrete by Concreteworks (concreteworks.com); page 100-101—Kitchen countertop by Concreteworks (concreteworks.com), Crystal Sphere Pendant by Alison Berger for Holly Hunt (hollyhunt.com), Catenary Bar Stools by Token (tokennyc.com), Custom Cabinets by Lloyd's Custom Woodwork (lcwoodwork.com)
Designer/Architect: John Maniscalco Architecture—San Francisco, CA and m-architecture.com; Shawback Design—Napa, California and shawbackdesign.com; Ethan Allen Construction—Canyon, California and ethanallenconstruction.com
Photographer: Mariko Reed—
marikoreed.com

18. Green Light Kitchen
Pages: 102-103
Place: Walls in a commercial kitchen in Marina del Rey, California
Date: 2013
Tile: Glossy green 3x6" field tile from Waterworks (waterworks.com)
Other Materials: Japanese yellow teapot from Tortoise General Store (tortoisegeneralstore.com)
Designer/Architect: DISC Interiors—Los Angeles, California and discinteriors.com
Photographer: Mariko Reed—
marikoreed.com

19. Parco dei Principi Hotel
Pages: 104-109
Place: Walls and floors in a hotel in Sorrento, Italy
Date: 1962
Tile: Floor tile is by Ceramica d'Agostino; page 104,105—Pebbles are by Ceramica Joo; page 107,108-109—Ceramic plaques are by Fausto Melotti
Designer/Architect: Giò Ponti
Photographer: Janos Grapow—hotelphotography.it; Cesare Naldi—cesarenaldi.com

20. Kogure House
Pages: 110-115
Place: Kitchen counters, backsplash, and floors in a private home in Tokyo, Japan
Date: 1998
Tile: Moroccan handmade tile, manufacturer unknown
Designer/Architect: Jiro Murofushi
Photographer: Mariko Reed—
marikoreed.com

21. MAD House
Pages: 116 -119
Place: Fireplace surround and front

entry in a private home in Vancouver, Canada
Date: 2014
Tile: Page 117—Mexican earthenware tile by Tierra y Fuego (tierrayfuego.com); page 118-119—Heath Ceramics stoneware Wide Hex in Bright Yellow, New Jade Porcelain, and Campari Red (heathceramics.com)
Other Materials: Page 116—Neutra House Numbers in aluminum by DWR (dwr.com)
Designer/Architect: Marianne Amodio Architecture Studio—Vancouver, Canada and maastudio.com
Photographer: Janis Nicolay—janisnicolay.com

22. Stark House
Pages: 120-121
Place: Fireplace surround in a private home in San Francisco, California
Date: 2011
Tile: Heath Ceramics stoneware 3x9"-Oval in Gunmetal (heathceramics.com)
Other Materials: Page 88—All furnishings are mid-century vintage
Designer/Architect: Jones | Haydu—San Francisco, California and joneshaydu.com
Photographer: Bruce Damonte—brucedamonte.com

23. White Brick House
Pages: 122-125
Place: Kitchen floors and walls in a private home in Portland, Oregon
Date: 2011
Tile: Page 122-123—Paccha Rings concrete floor tile by Popham Design (pophamdesign.com); page 122-125—Basix earthenware 2x4" Field Tile in White Gloss by Ken Mason Tile (kmt-bcia.com)
Other Materials: Page 124-125—Custom countertops are 25" wide walnut slabs, Fireclay Farmhouse sink in white by Franke (franke.com)
Designer/Architect: Jessica Helgerson Interior Design—Portland, Oregon and jhinteriordesign.com
Photographer: Mariko Reed—marikoreed.com

24. Waters House
Pages: 126-127
Place: Kitchen backsplash in a private home in Berkeley, California
Date: 2007
Tile: Page 92—Heath Ceramics stoneware field tile hand-cut to custom size in Lichen, Olive Gloss, Green Apple, and New Fawn; page 93—Heath Ceramics stoneware 2x4" Field
Tile in Lichen (heathceramics.com)
Other Materials: Concrete counters by Buddy Rhodes (buddyrhodes.com)
Designer/Architect: Alhorn/Hooven—Berkeley, California and alhornhooven.com
Photographer: Mariko Reed - marikoreed.com

25. Hillside House
Pages: 128-129
Place: Bathroom walls in a private home in Mill Valley, California
Date: 2011
Tile: Page 128—Heath Ceramics stoneware 2x4" Field Tile Dual Glaze in Stone Gray (heathceramics.com); page 129—Heath Ceramics stoneware 2x6" Field Tile in Stone White (heathceramics.com)
Other Materials: Page 129—Cast concrete sink and bath surround by Concreteworks (concreteworks.com)
Designer/Architect: SB Architects—San Francisco, California and SB-architects.com; Erin Martin Design—St. Helena, California and erinmartindesign.com
Photographer: Mariko Reed—marikoreed.com

26. McKenzie House
Pages: 130-131
Place: Kitchen wall in a private home in Borrego Springs, California
Date: 1980 with tile update in 2007
Tile: Heath Ceramics stoneware 3x9" Diamonds and Bowties in Frost (heathceramics.com)
Other Materials: Countertops are Corian
Designer/Architect: Maurice McKenzie; Stacey Chapman Paton
Photographer: JUCO—jucophoto.com

27. Komon Bath
Pages: 132-133
Place: Bathroom floors in a private home in Helsingør, Denmark
Date: 2011
Tile: Made a Mano Komon lava stone tiles (madeamano.com)
Designer/Architect: Made A Mano—Copenhagen, Denmark and madeamano.com
Photographer: Heidi Lerkenfeldt—lerkenfeldt.dk

28. Zamora Loft
Pages: 134-135
Place: Bathroom in a private home in Oakland, California
Date: 2011
Tile: Heath Ceramics stoneware Half Hex in Fog (heathceramics.com)
Designer/Architect: Christina Zamora—Oakland, California
Photographer: Jeffery Cross—jefferycross.com

29. Underwater Bath
Pages: 136-137
Place: Walls in a bathroom in a private home in San Francisco, California
Date: 2007
Tile: Glass tile by Bisazza (bisazza.com)

Designer/Architect: Envelope A+D—Berkeley, California and envelopead.com
Photographer: Todd Hido—toddhido.com

30. Yardhouse
Pages: 138-139
Place: Exterior walls on a commercial building in London, England
Date: 2014
Tile: Concrete tiles handmade on-site by Assemble Studio
Designer/Architect: Assemble Studio—London, England and assemblestudio.co.uk
Photographer: Assemble Studio

31. Llama Restaurant
Pages: 140-143
Place: Walls and floors of a restaurant in Copenhagen, Denmark
Date: 2014
Tile: Mexican cement tile by Original Mission tile (originalmissiontile.com)
Other Materials: Page 140, 142-143—Spine Barstools designed by Space Copenhagen for Fredericia (fredericia.com), Weight Here Candleholders designed by KiBiSi for Menu (menu.as), Mexican Beaded Skull by artist Santos Bautista (peyotepeople.com/blog/?p=133)
Designer/Architect: BIG—Copenhagen, Denmark and big.dk; Kilo Design—Copenhagen, DK and kilodesign.dk; Hz—Copenhagen, DK and h-z.dk
Photographer: Thomas Andersen

32. Roddick House
Pages: 144-147
Place: Kitchen walls, hallway floors and walls, and a fireplace in a private home in London, England
Date: 2009
Tile: Tiles by Emery & Cie (emeryetcie.com)
Other Materials: Page 144—Range Hood is Platinum by Elica (elica.com); page 146—Chair is Ernest Race, reproduced in a limited edition for Retrouvius (retrouvious.com).
Designer/Architect: Maria Speake, Retrouvius—London, England and retrouvious.com
Photographer: Morten Holtum—holtum.dk; styled by Lykke Foged—lykke-foged.dk

33. Pluijm House
Pages: 148-151
Place: Bathroom backsplash, kitchen backsplash and island in a private home in Marrakech, Morocco
Date: 2010
Tile: Tiles and many accessories by Ank van der Pluijm's own Household Hardware (householdhardware.nl)
Designer/Architect: Ank van der Pluijm—Amsterdam, Holland and householdhardware.nl
Photographer: Morten Holtum—holtum.dk; styled by Lykke Foged—lykke-foged.dk

34. Zittel House
Pages: 152-153
Place: Kitchen and dining room floors and walls in a private home in Joshua Tree, California
Date: 2010
Tile: Custom handmade 12x12" cement tiles produced by Arena México, Arte Contemporáneo (arenamexico.com)
Other Materials: Page 153—Dining Table by Andrea Zittel for Andrea Rosen Gallery (andrearosengallery.com), Eames fiberglass side chairs
Designer/Architect: Andrea Zittel—Joshua Tree, California and zittel.org
Photographer: Jessica Eckert—jessicaeckert.com, courtesy of the artist and Andrea Rosen Gallery, New York.

35. FINE Design Office
Pages: 154-155
Place: Floors and walls in an office in Portland, Oregon
Date: 2013
Tile: Heath Ceramics stoneware 3x9" and 2x9" Field Tile in Modern Blue and 2x9" Field Tile in Crystal Blue (heathceramics.com)
Other Materials: Island/Table by MADE (made-studio.com), Caravaggio pendant designed by Cecilie Manz for Lightyears (lightyears.dk)
Designer/Architect: Boora Architects—Portland, Oregon and boora.com
Photographer: Mariko Reed—marikoreed.com

36. Peaks View House
Pages: 156-157
Place: Bathroom walls and exterior BBQ wall in a private home in Wilson, Wyoming
Date: 2009
Tile: Page 156—Heath Ceramics stoneware 3x9" Field Tile in New Modern Blue (heathceramics.com); page 157—Heath Ceramics stoneware 3x12" Field Tile in New Redwood (heathceramics.com)
Other Materials: Page 156—Bathtub by Zuma (zumacollection.com), Bath fixtures hardware by Hansgrohe (hansgrohe.com)
Designer/Architect: Carney Logan Burke Architects—Jackson, Wyoming and clbarchitects.com
Photographer: Matthew Millman—matthewmillman.com

37. Margarido House
Pages: 158-159
Place: Shower walls and floors in a private home in Oakland, California
Date: 2010
Tile: Page 158—Heath Ceramics stoneware 2x6" Field Tile in New Seafoam and New Jade Porcelain

(heathceramics.com); page 159—Heath Ceramics stoneware 2x9" Field Tile in New Crystal Blue (heathceramics.com)
Other Materials: Page 159—Raindance shower head by Hans Grohe (hansgrohe.com), wide plank wood flooring in walnut, Eames Walnut Stool by Herman Miller (hermanmiller.com)
Designer/Architect: Onion Flats—Philadelphia, PA and onionflats.com; Medium Plenty—Oakland, California and mediumplenty.com
Photographer: Mariko Reed—marikoreed.com

38. São Paulo House
Pages: 160-163
Place: Dining room walls and exterior walls in a private home in São Paulo, Brazil
Date: 2012
Tile: Coated ceramic mosaic from Mazza Cerâmicas (mazzaceramicas.com.br)
Other Materials: Page 120—Cast concrete built custom onsite, Sofa and side table by Minotti (minotti.com), Wishbone dining chairs by Hans Wegner for Carl Henson & Søn, Discoco pendant lamp designed by Christophe Mathieu for Marset (marset.com), Taccia Table lamp by Achille and Pier Giacomo Castiglioni for Flos (flos.com)
Designer/Architect: Guilherme Torres—São Paulo, Brazil and guilhermetorres.com.br
Photographer: Denilson Machado—mcaestudio.com.br

39. Romita Comedor
Pages: 164-167
Place: Walls and floors in a restaurant in Mexico City, Mexico
Date: 2012
Tile: Page 164,165,167—Custom cement tiles designed by Richard Mozka and made by Rayito de Sol (unrayitodesol.com); page 166-167—Floor tiles are original to the building dating from the early 1900s
Designer/Architect: Rodrigo Espinoza, Marcela Lugo and Arturo Dib
Photographer: Mariko Reed—marikoreed.com; Jaime Navarro

40. Ace Hotel
Pages: 168-171
Place: Walls in the restaurant and bar areas of a hotel in Los Angeles, California
Date: 2014
Tile: Page 168,169—Slash Tile from the Sittio collection by Commune for Exquisite Surfaces (xsurfaces.com); page 170-171—Custom cast concrete tile
Other Materials: Page 170-171-Pencil Cedar Stool and Pencil Cedar Table with steel legs both by Alma Allen (almaallen.com), vintage Mexican Butaca Chair, felt wall hanging by Tanya Aguiniga (aguinigadesign.com)
Designer/Architect: Commune Design—Los Angeles, California and communedesign.com
Photographer: Mariko Reed—marikoreed.com

41. Avenue Loft
Pages: 172-173
Place: Kitchen walls in a private home in Portland, Oregon
Date: 2013
Tile: Glazed brick in white by Gran Brique (granbrique.com)
Other Materials: Kitchen counters are Carrara Marble, kitchen faucet by Rohl (rohlhome.com)
Designer/Architect: Jessica Helgerson Interior Design—Portland, Oregon and jhinteriordesign.com
Photographer: Mariko Reed—marikoreed.com

42. Bailey-Petravic House
Pages: 174-177
Place: Kitchen backsplash, fireplace surround, and bathroom walls in a private home in Sausalito, California
Date: 2005 and 2009
Tile: Page 174—Heath Ceramics stoneware 2x8"-Field Tile custom with divot in Bright Yellow (heathceramics.com); page 175,176—On the walls is Heath Ceramics stoneware Dandelion Seed tapestry pattern, on the floors is Heath Ceramics stoneware 4x8" Field Tile in Steam (heathceramics.com); page 176—Heath Ceramics stoneware 2x4" and 2x2"-Field Tile in Hematite and Gunmetal (heathceramics.com)
Other Materials: Page 175—Concrete counters by Concreteworks (concreteworks.com); page 176—Towel warmer by Runtal (runtal.com); page 177—Vintage fireplace by Fire Hood, ceramic pendant lamps by Adam Silverman (adamsilverman.net), felt rug by Peace Industry (peaceindustry.com), Cesca Armchair designed by Marcel Breuer for Knoll (knoll.com), Cross Extension Table designed by Matthew Hilton for Case (casefurniture.co.uk), all pottery by Heath Ceramics (heathceramics.com)
Designer/Architect: Barbara Brown—Sausalito, California and bbrownarchitect.com; Catherine Bailey
Photographer: Leslie Williamson—lesliewilliamson.com

43. Turner House

Pages: 178-179
Place: Bathroom wall and tub surround in a private home in Larkspur, California
Date: 2013
Tile: Penny Round Tiles in Chalk from the Savoy Mosaics collection by Ann Sacks (annsacks.com)
Other Materials: Bathtub is Klassik Duo by Kaldewei (kaldewei.com)
Designer/Architect: Jensen Architects—San Francisco, CA and jensen-architects.com; Nicole Hollis Interior Design—San Francisco, California and nicolehollis.com
Photographer: Mariko Reed—marikoreed.com

44. East Bay Hills House

Pages: 180-181
Place: Fireplace surround in a private home in El Cerrito, California
Date: 2006
Tile: Heath Ceramics stoneware 2x9" Field Tile in Gunmetal (heathceramics.com)
Other Materials: Page 180—House numbers by Custom House Numbers (customhousenumbers.com); page 181—Floors are custom colored concrete over radiant heating by Creative Spaces Builders (creativespacesbuilders.com)
Designer/Architect: Ohashi Design Studio—Emeryville, California and ohashidesign.com
Photographer: John Sutton—johnsuttonphotography.com

45. Hole in the Wall House

Pages: 182-185
Place: Attic space (in a former barn) in a private home in Föhr, Germany
Date: 2012
Tile: White square tile, custom drilled hole filled with blue cement.
Other Materials: page 184-185—Floors are natural pine wood, staircase screen is blue polypropylene rope
Designer/Architect: Francesco Di Gregorio—Parma, Italy and francescodigregorio.it; Karin Matz—Stockholm, Sweden and karinmatz.se
Photographer: Francesco Di Gregorio

46. Matbaren Bistro

Pages: 186-187
Place: Floors in a restaurant inside the Grand Hotel in Stockholm, Sweden
Date: 2007
Tile: Custom Encaustic Ceramic Tiles by Maw & Company (mawandco.com)
Other Materials: Page 186—Carimate red chairs designed by Vico Magistretti, Tivoli pendant lamps designed by Jørn Utzon; page 154—Storängen Chair designed by Mårten Cyrén for Pyra, Artichoke pendant lamp designed by Poul Henningsen for Louis Poulsen (louispoulsen.com)
Designer: Studioilse—London, England and studioilse.com
Photographer: Lisa Cohen—lisacohenphotography.com

47. Vintage House

Pages. 188-191
Place: Kitchen walls in a private home in Portland, Oregon
Date: 2013
TIle: Field Tiles in white with custom trim pieces are by Pratt & Larson Ceramics (prattandlarson.com)
Other Materials: Page 188—yellow sofa is vintage from 1stdibs (1stdibs.com); page 190-191—Chandelier is vintage from 1stdibs (1stdibs.com), other vintage furnishings
Designer/Architect: Jessica Helgerson Interior Design—Portland, Oregon and jhinteriordesign.com
Photographer: Mariko Reed—marikoreed.com

48. Lord House

Pages. 192-195
Place: Bathroom floors and kitchen walls in a private home in San Francisco, California
Date: 2014
TIle: Page 193,194—Heath Ceramics stoneware 5" Hex in Blue Fog and Shade Light (heathceramics.com); page 192,195—Heath Ceramics stoneware 3x6" Field Tile in Blue Fog (heathceramics.com)
Other Materials: Page 192—Kitchen sink faucet is Oxygene by Gessi (gessi.com); page 193—Clawfoot tub by Sunrise Speciality (sunrisespecialty.com), tub faucet by Watermark (watermark-designs.com); page 195—Corian counters with kitchen cabinets by Henry Built (henrybuilt.com)
Designer/Architect: Jeni Gamble—San Francisco, California and gambleplusdesign.com; Construction by Stingray Builders—San Francisco, California and stingraybuilders.com
Photographer: Mariko Reed—marikoreed.com

49. Neisha Crosland Studio

Pages. 196-199
Place: Interior and exterior walls, and exterior floors in a private studio in London, England
Date: 2012
Tile: Page 196, 197—Hand painted terracotta ceramic tiles from the Navajo collection designed by Neisha Crosland (neishacrosland.com) for De Ferranti (deferranti.com); page 198—Molly stenciled stone tiles designed by Neisha Crosland (neishacrosland.com) for De Ferranti (deferranti.com); page 199—Hand-painted Encaustic Cement tiles

from the Florentine collection designed by Neisha Crosland (neishacrosland.com) for Fired Earth (firedearth.com)
Other Materials: Page 196—Tio Chairs designed by Chris Martin for Massproductions (massproductions-online.com)
Designer/Architect: Greenway Architects—London, England and greenwayarchitects.co.uk; Garden by Sean Walters—Great Missenden, England and theplantspecialist.co.uk
Photographer: Mariko Reed—marikoreed.com

50. Chiselhurst House
Pages. 200-201
Place: Fireplace surround and bathroom floor in a private home in Los Angeles, California
Date: 2011
Tile: Page 200—Heath Ceramics stoneware 6x12" Field Tile in Midnight #6 (heathceramics.com); page 201—Granada Tile cement tile 8x8" Vegas pattern in Mustard and White (granadatile.com)
Other Materials: Page 200—Furnishings are vintage finds by the homeowner; page 201—Empire bathtub by Waterworks (waterworks.com), Freestanding tub spout is .25 by Waterworks (waterworks.com)
Designer: Bestor Architecture—Los Angeles, California and bestorarchitecture.com
Photographer: Ray Kachatorian—kachatorian.com

51. Buena Vista House
Pages: 202-205
Place: Kitchen backsplash, shower, and fireplace in a private residence in San Francisco, California
Date: 2014
Tile: Page 202—Heath Ceramics stoneware Wide Hex in Shade Light (heathceramics.com); page 203—Heath Ceramics stoneware 2x9" Field Tile in Blue Fog (heathceramics.com); page 204-205—Heath Ceramics stoneware Oval in New Chamois (heathceramics.com)
Other Materials: Page 204-205—three panel wall art by glass artist Jonah Ward (jonahward.com)
Designer/Architect: George Bradley | Architecture + Design—San Francisco, California and gabarch.com
Photographer: Mariko Reed—marikoreed.com

52. Wanzenberg House
Pages: 206-209
Place: Fireplace surround, kitchen and laundry room walls in private homes in New York, New York and Taghkanic, New York
Date: 2014
Tile: Page 206—Heath Ceramics stoneware 2x8" Field Tile in Bright Yellow (heathceramics.com); page 207—Heath Ceramics stoneware 2x2", 2x4", 2x6", 2x9", 3x1", 3x3", 3x6", 3x9" field tile in Campari Red, Pomegranate, and two little tiles in Yellowstone (heathceramics.com); page 208-209—Heath Ceramics stoneware 3x9" Ovals in Volcano and set in bands of painted wood and matching grout (heathceramics.com)
Other Materials: Page 206—Laundry sink is a J.L. Mott Iron Works restored antique, ceramic pendant lamp is vintage designed by Lee Rosen for Design Technics, ceramic wall plaques (also on page 208) are vintage by Francesca Mascitti Lindh for Arabia (arabia.fi); page 207—Chairs and round table are custom designed by Alan Wanzenberg (alanwanzenberg.com), ceramic pieces on mantle are 1950s vintage pottery from the French village of La Born; page 208-209—Kitchen countertop is Olympian White Vermont Danby, cabinetry wood is White Rock Maple
Designer/Architect: Alan Wanzenberg—New York, New York and alanwanzenberg.com
Photographer: Michelle Rose Studio—michellerosestudio.com

INDEX